KT-228-764

UNIT 1

FORCES AND MOTION

UNIT 2

ELECTRICITY

UNIT 3

WAVES

UNIT 4

ENERGY RESOURCES AND ENERGY TRANSFER

EDEXCEL INTERNATIONAL GCSE (9–1)

PHYSICS

Student Book

Brian Arnold

Penny Johnson

Steve Woolley

Published by Pearson Education Limited, 80 Strand, London, WC2R 0RL.

www.pearsonglobalschools.com

Copies of official specifications for all Edexcel qualifications may be found on the website: https://qualifications.pearson.com

Text © Pearson Education Limited 2017
Edited by Jane Read
Designed by Cobalt id
Typeset by TechSet Ltd
Original illustrations © Pearson Education Limited 2017
Illustrated by © TechSet Ltd
Cover design by Pearson Education Limited
Cover photo/illustration © ATELIER BRÜCKNER GMBH: Michael Jungblut

The rights of Brian Arnold, Penny Johnson and Steve Woolley to be identified as authors of this work have been asserted by them in accordance with the Copyright, Designs and Patents Act 1988.

First published 2017

20 19 18 17
10 9 8 7 6 5 4 3 2 1

British Library Cataloguing in Publication Data
A catalogue record for this book is available from the British Library

ISBN 978 0 435 18527 5

Printed by L.E.G.O. S.p.A. in Italy

Endorsement Statement
In order to ensure that this resource offers high-quality support for the associated Pearson qualification, it has been through a review process by the awarding body. This process confirms that this resource fully covers the teaching and learning content of the specification or part of a specification at which it is aimed. It also confirms that it demonstrates an appropriate balance between the development of subject skills, knowledge and understanding, in addition to preparation for assessment.

Endorsement does not cover any guidance on assessment activities or processes (e.g. practice questions or advice on how to answer assessment questions), included in the resource nor does it prescribe any particular approach to the teaching or delivery of a related course.

While the publishers have made every attempt to ensure that advice on the qualification and its assessment is accurate, the official specification and associated assessment guidance materials are the only authoritative source of information and should always be referred to for definitive guidance.

Pearson examiners have not contributed to any sections in this resource relevant to examination papers for which they have responsibility.

Examiners will not use endorsed resources as a source of material for any assessment set by Pearson. Endorsement of a resource does not mean that the resource is required to achieve this Pearson qualification, nor does it mean that it is the only suitable material available to support the qualification, and any resource lists produced by the awarding body shall include this and other appropriate resources.

Picture Credits
The publisher would like to thank the following for their kind permission to reproduce their photographs:

(Key: b-bottom; c-centre; l-left; r-right; t-top)

123RF.com: choneschones 63cr, cristimatei 262tl, Feliks Gurevich 80bl, homestudio 73, leungchopan 58, Melinda Nagy 18tc;
Alamy Stock Photo: age fotostock 28tr, 230, Ageev Rostislav 142, Alison Eckett 210, Ange 243cr, Archive Pics 233bl, Art Directors & TRIP 259, Ashley Cooper 163bl, Authentic Creations 120tl, Charles Stirling / Alamy Stock Photo 18tl, Chris Cooper-Smith 23tl, Chris Rose 110tr, Cultura Creative (RF) 212, Design Pics Inc 173tr, Digital Image Library 155cl, Dinodia Photos 48tl, DOE Photo 164, Emmanuel Lacoste 176cl, epa european pressphoto agency b.v. 87, Flake 60tl, frans lemmens 144cr, Fred Olivier / Nature Picture Library 148, GL Archive 233br, Granger Historical Picture Archive 34, Henry Westheim Photography 133tl, Horizon International Images Limited 143, 261, Image Source 75, ImagineThat 111cl, INTERFOTO 177, Jeff Rotman 173tc, John Joannides 139, Linda Richards 22tr, Andrew Michael 97tr, Michele Burgess 3tr, NASA 265, NASA / S.Dupuis 44, NASA Archive 37tr, The Natural History Museum 181tr, Nick Greening 3tl, paul ridsdale pictures 197, Peterz pics 158, Phanie 128br, philipus 82, Pictorial Press Ltd 150, 160, 187, 220, pixel shepherd 63tl, PjrStudio 110br, Pulsar Images 248tl, Radoslav Radev 120cl, Richard Wainscoat 262br, Robert Estall photo agency 162tl, samart boonyang 206, sciencephotos 79cr, 80, 103cl, Scott Ramsey 18tr, Sean Pavone 133tr, Studioshots 63b, Trevor Chriss 59, Alvin Wong 214tl, Владимир Галкин 9tc;
Fotolia.com: annavaczi 144cl, artush 32, izzzy71 144bl, ktsdesign 106, schankz 176cr, Szasz-Fabian Jozsef 9tr, troninphoto 23bl;
Getty Images: Aleksandrs Podskocijs / EyeEm 140, Carlos Herrera / Icon Sportswire / Corbis 133tc, Eco Images 165, Graiki 36, mevans 21bl, Michael Steele 279, michalz86 / iStock 271tr, Peter Turnley 28tl, ROBERT SULLIVAN / AFP 110cl, Science & Society Picture Library 155bl, 163cl, Scott Eells / Bloomberg 113bl, ViewStock 21tl, YinYang 22tl;
Maritime & Coastguard Agency: 109cr;
Pearson Education Ltd: Studio 8 71, Gareth Boden 79tl, Coleman Yuen. Pearson Education Asia Ltd 60bl, Jules Selmes 68, 144br, Tsz-shan Kwok 61;
Science Photo Library Ltd: Andrew Lambert Photography 12tr, 25, 188, CNRI 241, David Parker 153br, Dorling Kindersley / UIG 141, ESA - A. Le Floc'h 37tl, DR P. MARAZZI 242tr, Martyn F. Chillmaid 103tr, CORDELIA MOLLOY 20tl, 70tl, 70tr, PATRICK LANDMANN 248tr, R. MEGNA / FUNDAMENTAL PHOTOS 67, Sputnik 159tc, TONY & DAPHNE HALLAS 153cl, TRL Ltd. 45, WMAP SCIENCE TEAM, NASA 273;
Shutterstock.com: 805084 132, 111cr, Vartanov Anatoly 110tl, bikeriderlondon 109bl, Sylvie Bouchard 90, Brian Kinney 155tr, Cardens Design 258, Norman Chan 72, CreativeNature.nl 163bl, devy 9tl, Dja65 89, 271bl, emel82 162cl, EpicStockMedia 96, Martin Fischer 84, Galushko Sergey 128cl, imagedb.com 62, Joanne Harris and Daniel Bubnich 121, karrapavan 196, MidoSemsem 120cr, MIGUEL GARCIA SAAVEDRA 60br, Minerva Studio 179, morchella 123, NADA GIRL 91, Nicholas Sutcliffe 126, Noraluca013 159tl, Pavel L Photo and Video 113tr, portumen 163cr, Przemyslaw Skibinski 115, punksid 267, ra3rn 204, ronstik 236, Sander van Sinttruye 266tl, Santi Rodriguez 128cr, Olga Selyutina 97br, SFC 102, Smileus 100, Stanislav Tiplyashin 161, STUDIOMAX 145, tcly 48tr, Tiago Ladeira 147, topseller 173tl, valdis torms 40, Valeriy Lebedev 266bl, Villiers Steyn 128tr, Volodymyr Goinyk 172, vovan 159tr, Voyagerix 20bl, Wolfgang Kloehr 266tr, YanLev 5tl;
TWI Ltd. Cambridge: 243bl

Cover images: *Front:* ATELIER BRÜCKNER GMBH: Michael Jungblut

All other images © Pearson Education

Disclaimer: neither Pearson, Edexcel nor the authors take responsibility for the safety of any activity. Before doing any practical activity you are legally required to carry out your own risk assessment. In particular, any local rules issued by your employer must be obeyed, regardless of what is recommended in this resource. Where students are required to write their own risk assessments they must always be checked by the teacher and revised, as necessary, to cover any issues the students may have overlooked. The teacher should always have the final control as to how the practical is conducted.

UNIT 5

SOLIDS, LIQUIDS AND GASES

UNIT 6

MAGNETISM AND ELECTROMAGNETISM

UNIT 7

RADIOACTIVITY AND PARTICLES

UNIT 8

ASTROPHYSICS

PHYSICS ONLY

ABOUT THIS BOOK

This book is written for students following the Edexcel International GCSE (9–1) Physics specification and the Edexcel International GCSE (9–1) Science Double Award specification. You will need to study all of the content in this book for your Physics examination. However, you will only need to study some of it if you are taking the Double Award specification. The book clearly indicates which content is in the Physics examination and not in the Double Award specification. To complete the Double Award course you will also need to study the Biology and Chemistry parts of the course.

In each unit of this book, there are concise explanations and worked examples, plus numerous exercises that will help you build up confidence. The book also describes the methods for carrying out all of the required practicals.

The language throughout this textbook is graded for speakers of English as an additional language (EAL), with advanced Physics specific terminology highlighted and defined in the glossary at the back of the book. A list of command words, also at the back of the book, will help you to learn the language you will need in your examination.

You will also find that questions in this book have Progression icons and Skills tags. The Progression icons refer to Pearson's Progression scale. This scale – from 1 to 12 – tells you what level you have reached in your learning and will help you to see what you need to do to progress to the next level. Furthermore, Edexcel have developed a Skills grid showing the skills you will practise throughout your time on the course. The skills in the grid have been matched to questions in this book to help you see which skills you are developing. You can find Pearson's Progression scale and Edexcel's Skills grid at www.pearsonschoolsandfecolleges.co.uk along with guidelines on how to use them.

Learning Objectives show what you will learn in each Chapter.

Units boxes tell you which units – for example, metres, grams and seconds – you will need to know and use for the study of a topic.

Physics Only sections show the content that is on the Physics specification only and not the Double Award specification. All other content in this book applies to Double Award students.

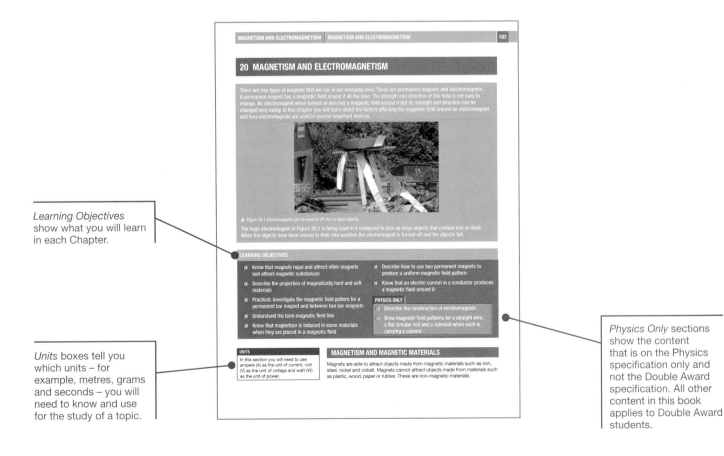

Key Point boxes summarise the essentials.

Extension Work boxes include content that is not on the specification and which you do not have to learn for your examination. However, the content will help to extend your understanding of the topic.

Hint boxes give you tips on important points to remember in your examination.

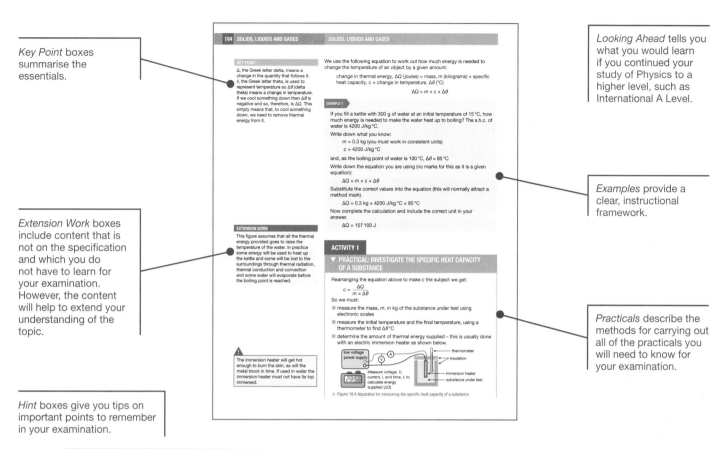

Looking Ahead tells you what you would learn if you continued your study of Physics to a higher level, such as International A Level.

Examples provide a clear, instructional framework.

Practicals describe the methods for carrying out all of the practicals you will need to know for your examination.

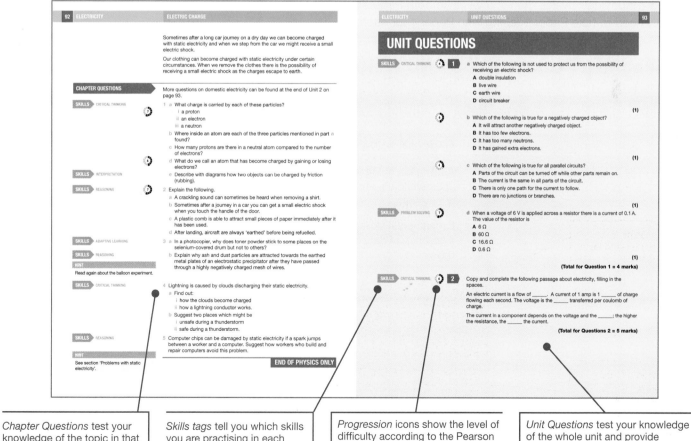

Chapter Questions test your knowledge of the topic in that chapter.

Skills tags tell you which skills you are practising in each question.

Progression icons show the level of difficulty according to the Pearson International GCSE Science Progression Scale.

Unit Questions test your knowledge of the whole unit and provide quick, effective feedback on your progress.

ASSESSMENT OVERVIEW

The following tables give an overview of the assessment for this course.

We recommend that you study this information closely to help ensure that you are fully prepared for this course and know exactly what to expect in the assessment.

PAPER 1	SPECIFICATION	PERCENTAGE	MARK	TIME	AVAILABILITY
Written examination paper Paper code 4PH1/1P and 4SD0/1P Externally set and assessed by Edexcel	Physics Double Award	61.1%	110	2 hours	January and June examination series First assessment June 2019

PAPER 2	SPECIFICATION	PERCENTAGE	MARK	TIME	AVAILABILITY
Written examination paper Paper code 4PH1/2P Externally set and assessed by Edexcel	Physics	38.9%	70	1 hour 15 mins	January and June examination series First assessment June 2019

If you are studying Physics then you will take both Papers 1 and 2. If you are studying Science Double Award then you will only need to take Paper 1 (along with Paper 1 for each of the Biology and Chemistry courses).

ASSESSMENT OBJECTIVES AND WEIGHTINGS

ASSESSMENT OBJECTIVE	DESCRIPTION	% IN INTERNATIONAL GCSE
AO1	Knowledge and understanding of physics	38%–42%
AO2	Application of knowledge and understanding, analysis and evaluation of physics	38%–42%
AO3	Experimental skills, analysis and evaluation of data and methods in physics	19%–21%

EXPERIMENTAL SKILLS

In the assessment of experimental skills, students may be tested on their ability to:

- solve problems set in a practical context
- apply scientific knowledge and understanding in questions with a practical context
- devise and plan investigations, using scientific knowledge and understanding when selecting appropriate techniques
- demonstrate or describe appropriate experimental and investigative methods, including safe and skilful practical techniques
- make observations and measurements with appropriate precision, record these methodically and present them in appropriate ways
- identify independent, dependent and control variables
- use scientific knowledge and understanding to analyse and interpret data to draw conclusions from experimental activities that are consistent with the evidence
- communicate the findings from experimental activities, using appropriate technical language, relevant calculations and graphs
- assess the reliability of an experimental activity
- evaluate data and methods taking into account factors that affect accuracy and validity.

CALCULATORS

Students are permitted to take a suitable calculator into the examinations. Calculators with QWERTY keyboards or that can retrieve text or formulae will not be permitted.

UNIT 1
FORCES AND MOTION

Forces make things move, like this Atlas V rocket carrying the Cygnus spacecraft up to the International Space Station. Forces hold the particles of matter together and keep us on the Earth. Forces can make things slow down. This is useful when we apply the brakes when driving a car! Forces can change the shape of things, sometimes temporarily and sometimes permanently. Forces make things rotate and change direction.

1 MOVEMENT AND POSITION

It is very useful to be able to make predictions about the way moving objects behave. In this chapter you will learn about some equations of motion that can be used to calculate the speed and acceleration of objects, and the distances they travel in a certain time.

▲ Figure 1.1 The world is full of speeding objects.

LEARNING OBJECTIVES

- Plot and explain distance–time graphs

- Know and use the relationship between average speed, distance moved and time taken:

$$\text{average speed} = \frac{\text{distance moved}}{\text{time taken}}$$

- Practical: investigate the motion of everyday objects such as toy cars or tennis balls

- Know and use the relationship between acceleration, change in velocity and time taken:

$$\text{acceleration} = \frac{\text{change in velocity}}{\text{time taken}} \qquad a = \frac{(v-u)}{t}$$

- Plot and explain velocity–time graphs

- Determine acceleration from the gradient of a velocity–time graph

- Determine the distance travelled from the area between a velocity–time graph and the time axis

- Use the relationship between final speed, initial speed, acceleration and distance moved:

$(\text{final speed})^2 = (\text{initial speed})^2 + (2 \times \text{acceleration} \times \text{distance moved})$

$$v^2 = u^2 + (2 \times a \times s)$$

KEY POINT

Sometimes average speed is shown by the symbols v_{average} or $\bar{v}$ but in this book v will be used.

UNITS

PHYSICS ONLY

- torque (turning effect): newton metre (N m)
- momentum: kilogram metre per second (kg m/s).

In this section you will need to use kilogram (kg) as the unit of mass, metre (m) as the unit of length, and second (s) as the unit of time, You will find measurements of mass made in subdivisions of the kilogram, like grams (g) and milligrams (mg), measurements of length in multiples of the metre, like the kilometre (km), and subdivisions like the centimetre (cm) and millimetre (mm). You will also be familiar with other units for time: minutes, hours, days and years etc. You will need to take care to convert units in calculations to the base units of kg, m and s when you meet these subdivisions and multiples.

Other units come from these base units. In the first chapter you will meet the units for:
- speed and velocity: metre per second (m/s)
- acceleration: metre per second squared (m/s²).

In later chapters you will meet the units for:
- force: newton (N)
- gravitational field strength: newton per kilogram (N/kg)

Speed is a term that is often used in everyday life. Action films often feature high-speed chases. Speed is a cause of fatal accidents on the road. Sprinters aim for greater speed in competition with other athletes. Rockets must reach a high enough speed to put communications satellites in orbit around the Earth. This chapter will explain how speed is defined and measured and how distance–time graphs are used to show the movement of an object as time passes. We shall then look at changing speed – acceleration and deceleration. We shall use velocity–time graphs to find the acceleration of an object. We shall also find how far an object has travelled using its velocity–time graph. You will find out about the difference between speed and velocity on page 6.

AVERAGE SPEED

KEY POINT

Sometimes you may see 'd' used as the symbol for distance travelled, but in this book 's' will be used to be consistent with the symbol used in A level maths and physics.

A car travels 100 kilometres in 2 hours so the average speed of the car is 50 km/h. You can work this out by doing a simple calculation using the following definition of speed:

$$\text{average speed, } v = \frac{\text{distance moved, } s}{\text{time taken, } t}$$

$$v = \frac{s}{t}$$

The average speed of the car during the journey is the total distance travelled, divided by the time taken for the journey. If you look at the speedometer in a car you will see that the speed of the car changes from instant to instant as the accelerator or brake is used. The speedometer therefore shows the instantaneous speed of the car.

UNITS OF SPEED

Typically the distance moved is measured in metres and time taken in seconds, so the speed is in metres per second (m/s). Other units can be used for speed, such as kilometres per hour (km/h), or centimetres per second (cm/s). In physics the units we use are metric, but you can measure speed in miles per hour (mph). Many cars show speed in both mph and kilometres per hour (kph or km/h). Exam questions should be in metric units, so remember that m is the abbreviation for metres (and not miles).

REARRANGING THE SPEED EQUATION

If you are given information about speed and time taken, you will be expected to rearrange the speed equation to make the distance moved the subject:

$$\text{distance moved, } s = \text{average speed, } v \times \text{time, } t$$

and to make the time taken the subject if you are given the distance moved and speed:

$$\text{time taken, } t = \frac{\text{distance moved, } s}{\text{average speed, } v}$$

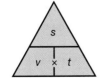

▲ Figure 1.2 You can use the triangle method for rearranging equations like $s = v \times t$.

REMINDER

To use the triangle method to rearrange an equation, cover up the part of the triangle that you want to find. For example, in Figure 1.2, if you want to work out how long (t) it takes to move a distance (s) at a given speed (v), covering t in Figure 1.2 leaves $\frac{s}{v}$, or distance divided by speed. If an examination question asks you to write out the equation for calculating speed, distance or time, always give the actual equation (such as $s = v \times t$). You may not get the mark if you just draw the triangle.

▲ Figure 1.3 A stopwatch will measure the time taken for the vehicle to travel the distance.

Suppose you want to find the speed of cars driving down your road. You may have seen the police using a mobile speed camera to check that drivers are keeping to the speed limit. Speed guns use microprocessors (computers on a 'chip') to produce an instant reading of the speed of a moving vehicle, but you can conduct a very simple experiment to measure car speed.

Measure the distance between two points along a straight section of road with a tape measure or 'click' wheel. Use a stopwatch to measure the time taken for a car to travel the measured distance. Figure 1.4 shows you how to operate your 'speed trap'.

1 Measure 50 m from a start point along the side of the road.

2 Start a stopwatch when your partner signals that the car is passing the start point.

3 Stop the stopwatch when the car passes you at the finish point.

▲ Figure 1.4 How to measure the speed of cars driving on the road

No measurements should be taken on the public road or pavement but it is possible to do so within the school boundary within sight of the road.

KEY POINT

You can convert a speed in m/s into a speed in km/h.

If the car travels 12.8 metres in one second it will travel

12.8 × 60 metres in 60 seconds (that is, one minute) and

12.8 × 60 × 60 metres in 60 minutes (that is, 1 hour), which is

46 080 metres in an hour or 46.1 km/h (to one decimal place).

We have multiplied by 3600 (60 × 60) to convert from m/s to m/h, then divided by 1000 to convert from m/h to km/h (as there are 1000 m in 1 km).

Rule: to convert m/s to km/h simply multiply by 3.6.

Using the measurements made with your speed trap, you can work out the speed of the car. Use the equation:

$$\text{average speed}, v = \frac{\text{distance moved}, s}{\text{time taken}, t}$$

So if the time measured is 3.9 s, the speed of the car in this experiment is:

$$\text{average speed}, v = \frac{50\,\text{m}}{3.9\,\text{s}}$$

$$= 12.8\,\text{m/s}$$

DISTANCE–TIME GRAPHS

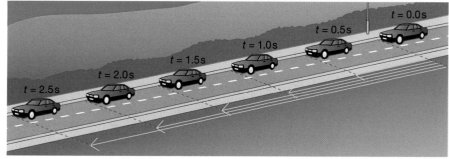

▲ Figure 1.5 A car travelling at constant speed

Figure 1.5 shows a car travelling along a road. It shows the car at 0.5 second intervals. The distances that the car has travelled from the start position after each 0.5 s time interval are marked on the picture. The picture provides a record of how far the car has travelled as time has passed. The table below shows the data for this car. You will be expected to plot a graph of the distance travelled (**vertical** axis) against time (horizontal axis) as shown in Figure 1.6.

Time from start/s	0.0	0.5	1.0	1.5	2.0	2.5
Distance travelled from start/m	0.0	6.0	12.0	18.0	24.0	30.0

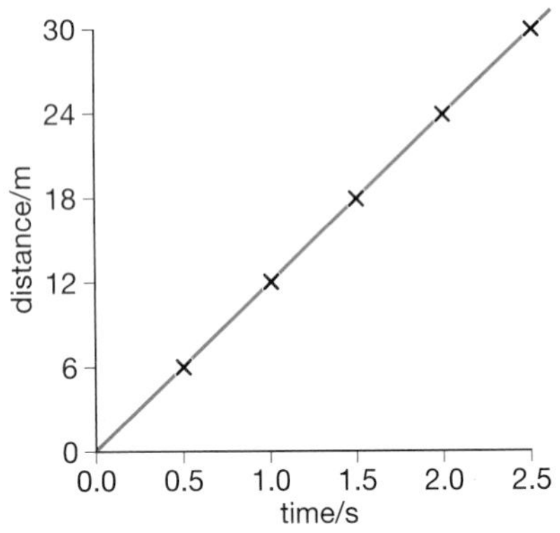

The distance–time graph tells us about how the car is travelling in a much more convenient form than the series of drawings in Figure 1.5. We can see that the car is travelling equal distances in equal time intervals – it is moving at a steady or constant speed. This fact is shown immediately by the fact that the graph is a straight line. The slope or **gradient** of the line tells us the speed of the car – the steeper the line the greater the speed of the car. So in this example:

$$\text{speed} = \text{gradient} = \frac{\text{distance}}{\text{time}} = \frac{30\,\text{m}}{2.5\,\text{s}} = 12\,\text{m/s}$$

▲ Figure 1.6 Distance–time graph for the travelling car in Figure 1.5

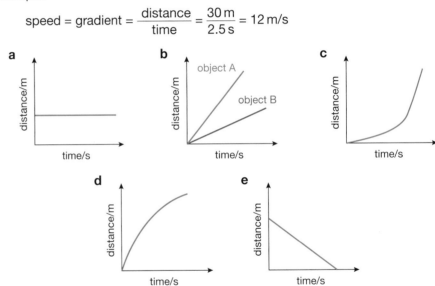

▲ Figure 1.7 Examples of distance–time graphs

In Figure 1.7a the distance is not changing with time – the line is horizontal. This means that the speed is zero. In Figure 1.7b the graph shows how two objects are moving. The red line is steeper than the blue line because object A is moving at a higher speed than object B. In Figure 1.7c the object is speeding up (**accelerating**) shown by the graph line getting steeper (gradient getting bigger). In Figure 1.7d the object is slowing down (decelerating).

KEY POINT

A curved line on distance–time graphs means that the speed or velocity of the object is changing. To find the speed at a particular instant of time we would draw a tangent to the curve at that instant and find the gradient of the tangent.

THE DIFFERENCE BETWEEN SPEED AND VELOCITY

Some displacement–time graphs look like the one shown in Figure 1.7e. It is a straight line, showing that the object is moving with constant speed, but the line is sloping down to the right rather than up to the right. The gradient of such a line is negative because the distance that the object is from the starting point is now decreasing – the object is going back on its path towards the start. **Displacement** means 'distance travelled in a particular direction' from a specified point. So if the object was originally travelling in a northerly direction, the negative gradient of the graph means that it is now travelling south.

KEY POINT

A vector is a quantity that has both size and direction. Displacement is distance travelled in a particular direction. **Force** is another example of a vector that you will meet in Chapter 2. The size of a force and the direction in which it acts are both important.

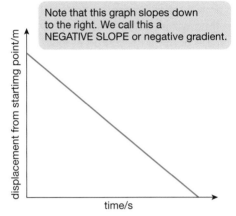

Note that this graph slopes down to the right. We call this a NEGATIVE SLOPE or negative gradient.

displacement from starting point/m

time/s

▲ Figure 1.8 In this graph displacement is decreasing with time.

KEY POINT

Always show your working when answering questions. You should show your working by putting the values given in the question into the equation.

Displacement is an example of a vector. Vector quantities have magnitude (size) and a specific direction.

Velocity is also a vector. Velocity is speed in a particular direction. If a car travels at 50 km/h around a bend, its speed is constant but its velocity will be changing for as long as the direction that the car is travelling in is changing.

$$\text{average velocity} = \frac{\text{increase in displacement}}{\text{time taken}}$$

EXAMPLE 1

The global positioning system (GPS) in Figure 1.9 shows two points on a journey. The second point is 3 km north-west of the first.

a A walker takes 45 minutes to travel from the first point to the second. Calculate the average velocity of the walker.

b Explain why the average speed of the walker must be greater than this.

a Write down what you know:

increase in displacement is 3 km north-west

time taken is 45 min (45 min = 0.75 h)

$$\text{average velocity} = \frac{\text{increase in displacement}}{\text{time taken}}$$

$$= 4 \text{ km/h north-west}$$

b The walker has to follow the roads, so the distance walked is greater than the straight-line distance between A and B (the displacement). The walker's average speed (calculated using distance) must be greater than his average velocity (calculated using displacement).

◀ Figure 1.9 The screen of a global positioning system (GPS). A GPS is an aid to navigation that uses orbiting satellites to locate its position on the Earth's surface.

ACTIVITY 1

▼ **PRACTICAL: INVESTIGATE THE MOTION OF EVERYDAY OBJECTS SUCH AS TOY CARS OR TENNIS BALLS**

You can use the following simple **apparatus** to investigate the motion of a toy car.

You could use this to measure the average speed, *v* of the car for different values of *h*.

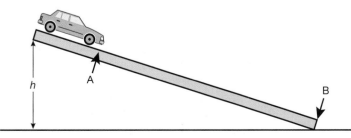

▲ Figure 1.10 Investigating how a toy car rolls down a slope

Heavy wooden runways need to be stacked and moved carefully. They are best used at low level rather than being placed on benches or tables where they may fall off. If heavy trolleys are used as 'vehicles', a 'catch box' filled with bubble wrap or similar material should be placed at the end of the runway.

You need to measure the height, h, of the raised end of the wooden track. The track must be securely clamped at the height under test and h should be measured with a metre rule making sure that the rule is **perpendicular** to the bench surface. Make sure that you always measure to the same point or mark on the raised end of the track (a fiducial mark).

To find the average speed you will use the equation:

$$\text{average speed, } v = \frac{\text{distance moved, } s}{\text{time taken, } t}$$

so you will need to measure the distance AB with a metre rule and measure the time it takes for the car to travel this distance with a stop clock. When timing with a stop clock, human **reaction** time will introduce measurement errors. To make these smaller the time to travel distance AB should, for a given value of h, be measured at least three times and an average value found. Always start the car from the same point, A. If one value is quite different from the others it should be treated as anomalous (the result is not accurate) and ignored or repeated.

The results should be presented in a table like the one below.

> You do not need to include these equations in your table headings but you may be asked to show how you did the calculations.

Distance/m	AB:

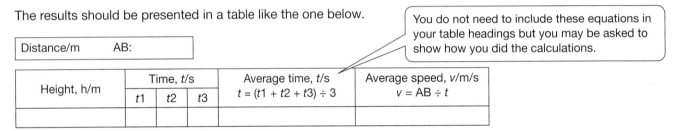

Height, h/m	Time, t/s			Average time, t/s $t = (t1 + t2 + t3) \div 3$	Average speed, v/m/s $v = AB \div t$
	t1	t2	t3		

In a question you may be given a complete set of results or you may be required to complete the table by doing the necessary calculations. You may be asked to plot a graph (see general notes above) and then draw a conclusion. The conclusion you draw must be explained with reference to the graph, for example, if the best fit line through the plotted points is a straight line and it passes through the origin (the 0, 0 point) you can conclude that there is a **proportional** relationship between the quantities you have plotted on the graph.

Some alternative methods

You could investigate the motion of moving objects using photographic methods either by:

- carrying out the experiment in a darkened room using a **stroboscope** to light up the object at regular known intervals (found from the **frequency** setting on the stroboscope) with the camera adjusted so that the shutter is open for the duration of the movement, or
- using a video camera and noting how far the object has travelled between each frame – the frame rate will allow you to calculate the time between each image.

In either case a clearly marked measuring scale should be visible.

Or you could use an electronically operated stop clock and electronic timing gates. This will let you measure the time that it takes for the moving object to travel over a measured distance. This has the advantage of removing timing errors produced by human reaction time.

You can also use timing gates to measure how the speed of the object changes as it moves.

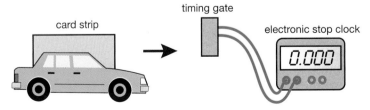

▲ Figure 1.11 Using a timing gate is a more accurate method for measuring time taken to travel a distance.

In this arrangement the stop clock will time while the card strip attached to the moving car passes through the timing gate. Measuring the length of the card strip and the time it takes for the card strip to pass through the timing gate allows you to calculate the average speed of the car as it passes through the timing gate.

ACCELERATION

Figure 1.12 shows some objects whose speed is changing. The plane must accelerate to reach take-off speed. In ice hockey, the puck (small disc that the player hits) decelerates only very slowly when it slides across the ice. When the egg hits the ground it is forced to decelerate (decrease its speed) very rapidly. Rapid deceleration can have destructive results.

▲ Figure 1.12 Acceleration ...

... constant speed ...

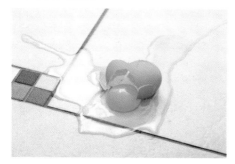

... and deceleration

Acceleration is the rate at which objects change their velocity. It is defined as follows:

$$\text{acceleration, } a = \frac{\text{change in velocity}}{\text{time taken, } t} \text{ or } \frac{\text{final velocity, } v - \text{initial velocity, } u}{\text{time taken, } t}$$

$$a = \frac{(v - u)}{t}$$

Why u? Simply because it comes before v!

Acceleration, like velocity, is a vector because the direction in which the acceleration occurs is important as well as the size of the acceleration.

UNITS OF ACCELERATION

Velocity is measured in m/s, so increase in velocity is also measured in m/s. Acceleration, the rate of increase in velocity with time, is therefore measured in m/s/s (read as 'metres per second per second'). We normally write this as m/s^2 (read as 'metres per second **squared**'). Other units may be used – for example, cm/s^2.

EXAMPLE 2

A car is travelling at 20 m/s. It accelerates steadily for 5 s, after which time it is travelling at 30 m/s. Calculate its acceleration.

Write down what you know:

initial or starting velocity, u = 20 m/s

final velocity, v = 30 m/s

time taken, t = 5 s

$$a = \frac{(v - u)}{t}$$

$$= \frac{30\,\text{m/s} - 20\,\text{m/s}}{5\,\text{s}}$$

$$= \frac{10\,\text{m/s}}{5\,\text{s}}$$

The car is accelerating at $2\,m/s^2$.

HINT

It is good practice to include units in equations – this will help you to supply the answer with the correct unit.

DECELERATION

Deceleration means slowing down. This means that a decelerating object will have a smaller final velocity than its starting velocity. If you use the equation for finding the acceleration of an object that is slowing down, the answer will have a negative sign. A negative acceleration simply means deceleration.

EXAMPLE 3

An object hits the ground travelling at 40 m/s. It is brought to rest in 0.02 s. What is its acceleration?

Write down what you know:

initial velocity, $u = 40$ m/s

final velocity, $v = 0$ m/s

time taken, $t = 0.02$ s

$$a = \frac{(v - u)}{t}$$
$$= \frac{0\,\text{m/s} - 40\,\text{m/s}}{0.02\,\text{s}}$$
$$= \frac{-40\,\text{m/s}}{0.02\,\text{s}}$$
$$= -2000\,\text{m/s}^2$$

In Example 3, we would say that the object is decelerating at 2000 m/s^2. This is a very large deceleration. Later, in Chapter 3, we shall discuss the consequences of such a rapid deceleration!

MEASURING ACCELERATION

EXTENSION WORK

Galileo was an Italian scientist who was born in 1564. He developed a telescope, which he used to study the movement of the planets and stars. He also carried out many experiments on motion (movement).

EXTENSION WORK

Though Galileo did not have a clock or watch (let alone an electronic timer), he used his pulse (the sound of his heart) and a type of water clock to achieve timings that were accurate enough for his experiments.

When a ball is rolled down a slope it is clear that its speed increases as it rolls – that is, it accelerates. Galileo was interested in how and why objects, like the ball rolling down a slope, speed up, and he created an interesting experiment to learn more about acceleration. A version of his experiment is shown in Figure 1.13.

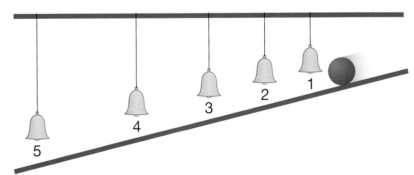

▲ Figure 1.13 Galileo's experiment. A ball rolling down a slope, hitting small bells as it rolls

Galileo wanted to discover how the distance travelled by a ball depends on the time it has been rolling. In this version of the experiment, a ball rolling down a slope strikes a series of small bells as it rolls. By adjusting the positions of the bells carefully it is possible to make the bells ring at equal intervals of time as the ball passes. Galileo noticed that the distances travelled in equal time intervals increased, showing that the ball was travelling faster as time passed. Galileo did not have an accurate way of measuring time (there were no digital stopwatches in seventeenth-century Italy!) but it was possible to judge equal time intervals accurately simply by listening.

Galileo also noticed that the distance travelled by the ball increased in a predictable way. He showed that the rate of increase of speed was steady or uniform. We call this uniform acceleration. Most acceleration is non-uniform – that is, it changes from instant to instant – but we shall only deal with uniformly accelerated objects in this chapter.

VELOCITY–TIME GRAPHS

The table below shows the distances between the bells in an experiment such as Galileo's.

Bell	1	2	3	4	5
Time/s	0.5	1.0	1.5	2.0	2.5
Distance of bell from start/cm	3	12	27	48	75

We can calculate the average speed of the ball between each bell by working out the distance travelled between each bell, and the time it took to travel this distance. For the first bell:

$$\text{velocity, } v = \frac{\text{distance moved, } s}{\text{time taken, } t}$$

$$= \frac{3\,\text{cm}}{0.5\,\text{s}} = 6\,\text{cm/s}$$

This is the average velocity over the 0.5 second time interval, so if we plot it on a graph we should plot it in the middle of the interval, at 0.25 seconds.

Repeating the above calculation for all the results gives us the following table of results. We can use these results to draw a graph showing how the velocity of the ball is changing with time. The graph, shown in Figure 1.14, is called a velocity–time graph.

Time/s	0.25	0.75	1.25	1.75	2.25
Velocity/cm/s	6	18	30	42	54

KEY POINT

The equations of motion we have learned work for uniform or constant acceleration only – therefore for objects with velocity–time graphs that are straight lines.

The graph in Figure 1.14 is a straight line. This tells us that the velocity of the rolling ball is increasing by equal amounts in equal time periods. We say that the acceleration is uniform in this case.

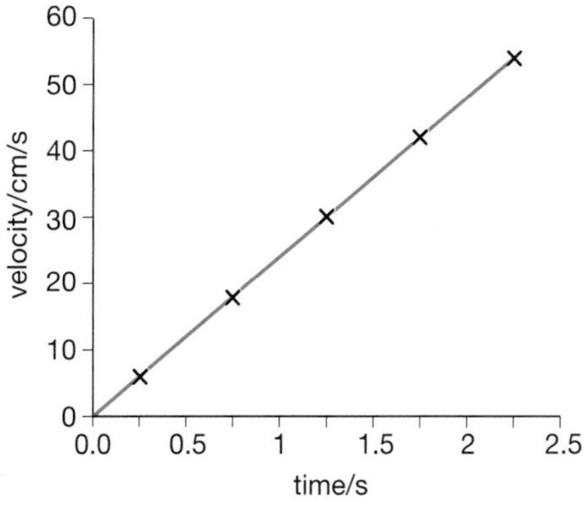

▲ Figure 1.14 Velocity–time graph for an experiment in which a ball is rolled down a slope. (Note that as we are plotting average velocity, the points are plotted in the middle of each successive 0.5 s time interval.)

A MODERN VERSION OF GALILEO'S EXPERIMENT

! A cylinder vacuum cleaner (or similar) used with the air-track should be placed on the floor as it may fall off a bench or stool. Also, beware of any trailing leads.

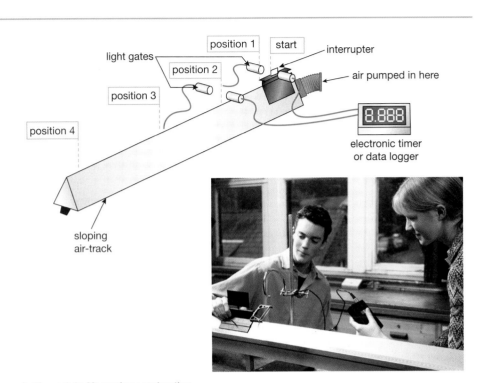

▲ Figure 1.15 Measuring acceleration

Today we can use data loggers to make accurate direct measurements that are collected and analysed by a computer. A spreadsheet program can be used to produce a velocity–time graph. Figure 1.15 shows a glider on a slightly sloping air-track. The air-track reduces **friction** because the glider rides on a cushion of air that is pushed continuously through holes along the air-track. As the glider accelerates down the sloping track the card stuck on it breaks a light beam, and the time that the glider takes to pass is measured electronically. If the length of the card is measured, and this is entered into the spreadsheet, the velocity of the glider can be calculated by the spreadsheet program using $v = \frac{s}{t}$.

Figure 1.16 shows velocity–time graphs for two experiments done using the air-track apparatus. In each experiment the track was given a different slope. The steeper the slope of the air-track the greater the glider's acceleration. This is clear from the graphs: the greater the acceleration the steeper the gradient of the graph.

The gradient of a velocity–time graph gives the acceleration.

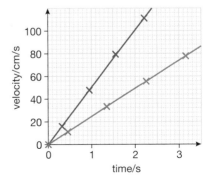

Air-track at 1.5°		Air-track at 3.0°	
Time/s	Av Vel. /cm/s	Time/s	Av Vel. /cm/s
0.00	0.0	0.00	0.0
0.45	11.1	0.32	15.9
1.35	33.3	0.95	47.6
2.25	55.6	1.56	79.4
3.15	77.8	2.21	111.1

▲ Figure 1.16 Results of two air-track experiments. (Note, once again, that because we are plotting average velocity in the velocity–time graphs, the points are plotted in the middle of each successive time interval – see page 11)

MORE ABOUT VELOCITY–TIME GRAPHS

GRADIENT

The results of the air-track experiments in Figure 1.16 show that the slope of the velocity–time graph depends on the acceleration of the glider. The slope or gradient of a velocity–time graph is found by dividing the increase in the velocity by the time taken for the increase, as shown in Figure 1.17. In this example an object is travelling at u m/s at the beginning and accelerates uniformly (at a constant rate) for t s. Its final velocity is v m/s. Increase in velocity divided by time is, you will recall, the definition of acceleration (see page 9), so we can measure the acceleration of an object by finding the slope of its velocity–time graph. The meaning of the slope or gradient of a velocity–time graph is summarised in Figure 1.17.

HINT

1 When finding the gradient of a graph, draw a big triangle.
2 Choose a convenient number of units for the length of the base of the triangle to make the division easier.

▲ Figure 1.17 Finding the gradient of a velocity–time graph

a shallow gradient – low acceleration   **b** steep gradient – high acceleration   **c** horizontal (zero gradient) – no acceleration   **d** negative gradient – negative acceleration (deceleration)

▲ Figure 1.18 The gradient of a velocity–time graph gives you information about the motion of an object at a glance.

AREA UNDER A VELOCITY–TIME GRAPH GIVES DISTANCE TRAVELLED

Figure 1.19a shows a velocity–time graph for an object that travels with a constant velocity of 5 m/s for 10 s. A simple calculation shows that in this time the object has travelled 50 m. This is equal to the shaded (coloured) area under the graph. Figure 1.19b shows a velocity–time graph for an object that has accelerated at a constant rate. Its average velocity during this time is given by:

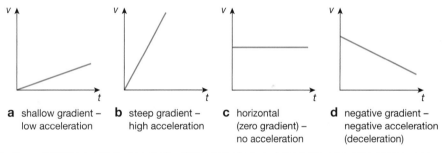

In this example the average velocity is, therefore:

$$\text{average velocity} = \frac{0\,\text{m/s} + 10\,\text{m/s}}{2}$$

which works out to be 5 m/s. If the object travels, on average, 5 metres in each second it will have travelled 20 metres in 4 seconds. Notice that this, too, is equal to the shaded area under the graph (given by the area equation for a triangle: area = $\frac{1}{2}$ base × height).

HINT

Find the distance travelled for more complicated velocity–time graphs by dividing the area beneath the graph line into rectangles and triangles. Take care that units on the velocity and time axes use the same units for time, for example, m/s and s, or km/h and h.

The area under a velocity–time graph is equal to the distance travelled by (displacement of) the object in a particular time interval.

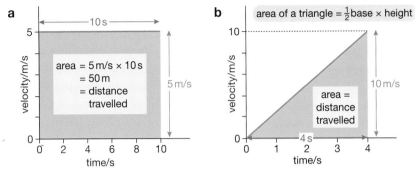

▲ Figure 1.19 a An object travelling at constant velocity; b An object accelerating at a constant rate

EQUATIONS OF UNIFORMLY ACCELERATED MOTION

You must remember the equation:

$$a = \frac{v - u}{t}$$

and be able to use it to calculate the acceleration of an object.

You may need to rearrange the equation to make another term the subject.

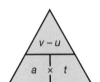 

▲ Figure 1.20 Cover $v - u$ to find $v - u = a \times t$

EXAMPLE 4

A stone accelerates from rest uniformly at 10 m/s² when it is dropped down a deep well. It hits the water at the bottom of the well after 5 s. Calculate how fast it is travelling when it hits the water.

You will need to make v the subject of this equation:

$$a = \frac{v - u}{t}$$

You can use the triangle method to show that $v - u = a \times t$

then add u to both sides of the equation to give:

$$v = u + at$$

(In words this tells you that the final velocity is the initial velocity plus the increase in velocity after accelerating for t seconds.)

State the things you have been told:

initial velocity, $u = 0$ m/s (It was stationary (standing still) at the start.)

acceleration, $a = 10$ m/s²

time, t, of the acceleration = 5 s

Substitute these into the equation: $v = 0$ m/s + (10 m/s² × 5 s)

Then calculate the result.

The stone hit the water travelling at 50 m/s (downwards).

You will also be required to use the following equation of uniformly accelerated motion:

(final speed)², v^2 = (initial speed)², u^2 + (2 × acceleration, a × distance moved, s)

$$v^2 = u^2 + 2as$$

EXAMPLE 5

A cylinder containing a vaccine is dropped from a helicopter hovering at a height of 200 m above the ground. The acceleration due to gravity is 10 m/s². Calculate the speed at which the cylinder will hit the ground.

You are given the acceleration, a = 10 m/s², and the distance, s = 200 m, through which the cylinder moves. The initial velocity, u, is not stated, but you assume it is 0 m/s as the helicopter is hovering (staying in one place in the air). Substitute these values in the given equation:

$$v^2 = u^2 + 2as$$

$$= 0 \text{ m/s}^2 + (2 \times 10 \text{ m/s}^2 \times 200 \text{ m})$$

$$= 4000 \text{ m}^2/\text{s}^2$$

therefore $v = \sqrt{(4000 \text{ m}^2/\text{s}^2)}$

$$= 63.25 \text{ m/s}$$

LOOKING AHEAD

The equations you have seen in this chapter are called the equations of uniformly accelerated motion. This means that they will give you correct answers when solving any problems that have objects moving with constant acceleration. In your exam you will only see problems where this is the case or very nearly so. Examples in which objects accelerate or decelerate (slow down) at a constant rate often have a constant acceleration due to the Earth's gravity (which we take as about 10 m/s²).

In real life, problems may not be quite so simple! Objects only fall with constant acceleration if we ignore air resistance and the distance that they fall is quite small.

These equations of uniformly accelerated motion are often called the 'suvat' equations, because they show how the terms s (distance moved), u (velocity at the start), v (velocity at the finish), a (acceleration) and t (time) are related.

CHAPTER QUESTIONS

More questions on speed and acceleration can be found at the end of Unit 1 on page 55.

SKILLS   PROBLEM SOLVING

1   A sprinter runs 100 metres in 12.5 seconds. Calculate the speed in m/s.

2   A jet can travel at 350 m/s. Calculate how far it will travel at this speed in:

   a   30 seconds

   b   5 minutes

   c   half an hour.

3   A snail crawls at a speed of 0.0004 m/s. How long will it take to climb a garden stick 1.6 m high?

4 Look at the following distance–time graphs of moving objects.

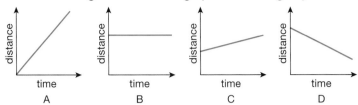

Identify in which graph the object is:

a moving backwards

b moving slowly

c moving quickly

d not moving at all.

5 Sketch a distance–time graph to show the motion of a person walking quickly, stopping for a moment, then continuing to walk slowly in the same direction.

6 Plot a distance–time graph using the data in the following table. Draw a line of best fit and use your graph to find the speed of the object concerned.

Distance/m	0.00	1.60	3.25	4.80	6.35	8.00	9.60
Time/s	0.00	0.05	0.10	0.15	0.20	0.25	0.30

7 The diagram below shows a trail of oil drips made by a car as it travels along a road. The oil is dripping from the car at a steady rate of one drip every 2.5 seconds.

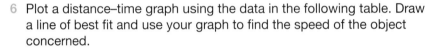

oil drips on the road

a Describe the way the car is moving.

b The distance between the first and the seventh drip is 135 metres. Determine the average speed of the car.

8 A car is travelling at 20 m/s. It accelerates uniformly at 3 m/s² for 5 s.

a Sketch a velocity–time graph for the car during the period that it is accelerating. Include numerical detail on the axes of your graph.

b Calculate the distance the car travels while it is accelerating.

9 Explain the difference between the following terms:

a average speed and instantaneous speed

b speed and velocity.

10 A sports car accelerates uniformly from rest to 24 m/s in 6 s. Calculate the acceleration of the car.

11 Sketch velocity–time graphs for an object:

a moving with a constant velocity of 6 m/s

b accelerating uniformly from rest at 2 m/s² for 10 s

c decelerating to rest at 4 m/s² for 5 s.

Include numbers and units on the velocity and time axes in each case.

SKILLS   PROBLEM SOLVING

12 A plane starting from rest accelerates at 3 m/s² for 25 s. Calculate the increase in velocity after:

a 1 s

b 5 s

c 25 s.

SKILLS   ANALYSIS

13 Look at the following sketches of velocity–time graphs of moving objects.

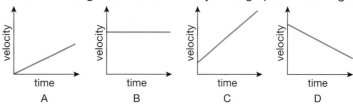

In which graph is the object:

a not accelerating

b accelerating from rest

c decelerating

d accelerating at the greatest rate?

SKILLS   INTERPRETATION

14 Sketch a velocity–time graph to show how the velocity of a car travelling along a straight road changes if it accelerates uniformly from rest for 5 s, travels at a constant velocity for 10 s, then brakes hard to come to rest in 2 s.

15 a Plot a velocity–time graph using the data in the following table:

Velocity/m/s	0.0	2.5	5.0	7.5	10.0	10.0	10.0	10.0	10.0	10.0
Time/s	0.0	1.0	2.0	3.0	4.0	5.0	6.0	7.0	8.0	9.0

Draw a line of best fit and use your graph to find:

b the acceleration during the first 4 s

c the distance travelled in:
 i the first 4 s of the motion shown
 ii the last 5 s of the motion shown

d the average speed during the 9 seconds of motion shown.

SKILLS   CRITICAL THINKING

16 The dripping car from Question 7 is still on the road! It is still dripping oil but now at a rate of one drop per second. The trail of drips is shown on the diagram below as the car travels from left to right.

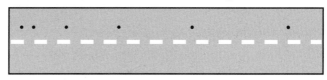

Describe the motion (the way the car is moving) using the information in this diagram.

SKILLS   PROBLEM SOLVING

17 This question uses the equation $v^2 = u^2 + 2as$.

a Explain what each of the terms in this equation represents.

b A ball is thrown vertically upwards at 25 m/s. Gravity causes the ball to decelerate at 10 m/s². Calculate the maximum height the ball will reach.

2 FORCES AND SHAPE

Forces are acting on us, and on objects all around us, all the time. In this chapter you will learn about different kinds of forces, how they may change the speed and direction of objects and how they can affect the shape of objects.

▲ Figure 2.1 Forces include pulling, falling due to gravity and squashing

LEARNING OBJECTIVES

- Describe the effects of forces between bodies such as changes in speed, shape or direction

- Identify different types of force such as gravitational or electrostatic

- Understand how vector quantities differ from scalar quantities

- Understand that force is a vector quantity

- Calculate the resultant force of forces that act along a line

- Know that friction is a force that opposes motion

- Practical: investigate how extension varies with applied force for helical springs, metal wires and rubber bands

- Know that the initial linear region of a force–extension graph is associated with Hooke's law

- Describe elastic behaviour as the ability of a material to recover its original shape after the forces causing deformation have been removed

Forces are simply pushes and pulls of one thing on another. Sometimes we can see their effects quite clearly. In Figure 2.1, the tug is pulling the tanker; the bungee jumper is being pulled to Earth by the force of gravity, and then (hopefully before meeting the ground) being pulled back up by the stretched elastic rope; the force applied by the crusher permanently changes the shape of the cars. In this chapter we will discuss different types of forces and look at their effects on the way that objects move.

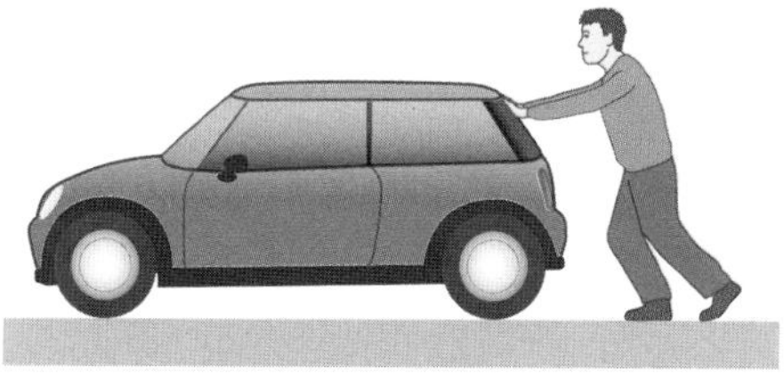

Figure 2.2 What forces do you think are working here?

ALL SORTS OF FORCES

If you are to study forces, first you need to notice them! As we have already said, sometimes they are easy to see and their effect is obvious. Look at Figure 2.2 and try to identify any forces that you think are involved.

You will immediately see that the man is applying a force to the car – he is pushing it. But there are quite a few more forces in the picture. To make the task a little easier we will limit our search to just those forces acting on the car. We will also ignore forces that are very small and therefore have little effect.

The man is clearly struggling to make the car move. This is because there is a force acting on the car trying to stop it moving. This is the force of friction between the moving parts in the car engine, gears, wheel axles and so on. This unhelpful force opposes the motion that the man is trying to achieve. However, when the car engine is doing the work to make the car go, the friction between the tyres and the road surface is vital. On an icy road even powerful cars may not move forward because there is not enough friction between the tyres and the ice.

Another force that acts on the car is the pull of the Earth. We call this a gravitational force or simply **weight**. If the car were to be pushed over the edge of a cliff, the effect of the gravitational force would be very clear as the car fell towards the sea. This leads us to realise that yet another force is acting on the car in Figure 2.2 – the road must be stopping the car from being pulled into the Earth. This force, which acts in an upward direction (going up) on the car, is called the reaction force. (A more complete name is **normal** reaction force. Here the word 'normal' means acting at 90° to the road surface.) All four forces that act on the car are shown in Figure 2.3.

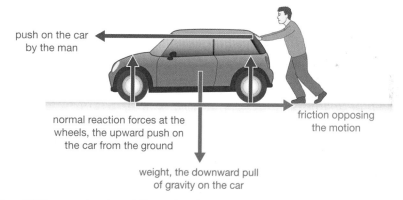

push on the car by the man

normal reaction forces at the wheels, the upward push on the car from the ground

friction opposing the motion

weight, the downward pull of gravity on the car

Figure 2.3 There are four types of force at work.

You will have realised by now that it is not just the size of the force that is important – the direction in which the force is acting is important, too.

Force is another example of a vector.

UNITS OF FORCE

The unit used to measure force is the newton (N), named after Sir Isaac Newton. Newton's study of forces is vital to our understanding of them today.

A force of one newton will make a **mass** of one kilogram accelerate at one metre per second squared.

This is explained more fully later (see Chapter 3). To give you an idea of the size of the newton, the force of gravity on a kilogram bag of sugar (its weight) is about 10 N; an average-sized apple weighs 1 N.

a

b

▲ Figure 2.4 More forces!

SOME OTHER EXAMPLES OF FORCES

It is not always easy to spot forces acting on objects. The compass needle in Figure 2.4a, which is a magnet, is affected by the magnetic force between it and the other magnet. Magnetic forces are used to make electric motors rotate, to hold fridge doors shut, and in many other situations.

If you comb your hair, you sometimes find that some of your hair sticks to the comb as shown in Figure 2.4b. This happens because of an electrostatic force between your hair and the comb. You can see a similar effect using a Van de Graaff generator, as shown in Figure 9.6 on page 87.

A parachute causes the parachutist to descend more slowly because an upward force acts on the parachute called air resistance or drag. Air resistance is like friction – it tries to oppose movement of objects through the air. Designers of cars, high-speed trains and other fast-moving objects try to reduce the effects of this force. Objects moving through liquids also experience a drag force – fast-moving animals that live in water have streamlined (smooth and efficient) shapes to reduce this force.

Hot air balloons are carried upwards in spite of the pull of gravity on them because of a force called upthrust. This is the upward push of the surrounding air on the balloon. An upthrust force also acts on objects in liquids.

More types of force, such as electric and nuclear forces, are mentioned in other chapters of this book. The rest of this chapter will look at the effects of forces.

MORE THAN ONE FORCE

As we saw earlier, in most situations there will be more than just one force acting on an object. Look at the man trying to push the car, shown in Figure 2.5. The two forces act along the same line, but in opposite directions. This means that one is negative (because it acts in the opposite direction to the other) and, if they are equal in size, they add up to zero and the car will not move.

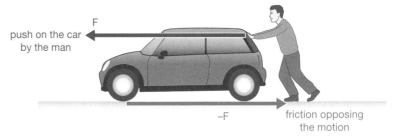

▲ Figure 2.5 The resultant force is zero because the two forces are balanced.

▲ Figure 2.6 The total pushing force is the sum of the two individual forces.

If the man gets someone to help him push the car, the forward force is bigger. Both of the forces pushing the car are acting in the same direction, so you can find the total forward force by adding the two forces together. If both people are pushing with a force of 300 N, then:

$$\text{total forward force} = 300 \text{ N} + 300 \text{ N}$$
$$= 600 \text{ N}$$

This means we can just add all the forces together to find the resultant force. As force is a vector quantity, we also need to think about the directions in which the forces are acting, and we do this by deciding which direction is the positive (+) direction. In this case, we can think of the force from the people as positive and the force from friction as negative. The + and – signs just show that the forces are acting in opposite directions.

So, if the force from friction is 300 N:

$$\textbf{unbalanced force} = 300 \text{ N} + 300 \text{ N} - 300 \text{ N}$$
$$= 300 \text{ N}$$

BALANCED AND UNBALANCED FORCES

Figure 2.7 shows two situations in which forces are acting on an object. In the tug of war contest the two teams are pulling on the rope in opposite directions. For much of the time the rope doesn't move because the two forces are balanced. This means that the forces are the same size but act in opposite directions along the line of the rope. There is no unbalanced force in one direction or the other. When the forces acting on something are balanced, the object does not change the way it is moving. In this case if the rope is stationary, it remains stationary. Eventually, one of the teams will become tired and its pull will be smaller than that of the other team. When the forces acting on the rope are unbalanced the rope will start to move in the direction of the greater force. There will be an unbalanced force in that direction. Unbalanced forces acting on an object cause it to change the way it is moving. The rope was stationary and the unbalanced forces acting on it caused it to accelerate.

The car in Figure 2.7 is designed to have an enormous acceleration from rest. As soon as it starts to move the forces that oppose motion – friction and drag – must be overcome. The **thrust** of the engine is, to start with, much greater than the friction and drag forces. This means that the forces acting on the car in the horizontal direction are unbalanced and the result is a change in the way that the car is moving – it accelerates! Once the friction forces balance the thrust the car no longer accelerates – it moves at a steady speed.

▲ Figure 2.7 Balanced forces and unbalanced forces

FRICTION

Friction is the force that causes moving objects to slow down and finally stop. The **kinetic energy** of the moving object is transferred to heat as work is done by the friction force. For the ice skater in Figure 2.8 the force of friction is very small so she is able to glide for long distances without having to do any work. It is also the force that allows a car's wheels to grip the road and make it accelerate – very quickly in the case of the racing cars in Figure 2.8.

Scientists have worked hard for many years to develop some materials that reduce friction and others that increase friction. Reducing friction means that machines work more efficiently (wasting less energy) and do not wear out so quickly. Increasing friction can help to make tyres that grip the road better and to make more effective brakes.

▲ Figure 2.8 The ice skater can glide because friction is low. The cars need friction to grip the road.

Friction occurs when solid objects rub against other solid objects and also when objects move through fluids (liquids and gases). Sprint cyclists and Olympic swimmers now wear special materials to reduce the effects of fluid friction so they can achieve faster times in their races. Sometimes fluid friction is very desirable – for example, when someone uses a parachute after jumping from a plane!

INVESTIGATING FRICTION

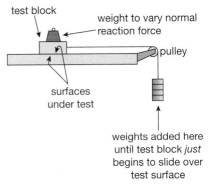

test block
weight to vary normal
reaction force
pulley
surfaces
under test
weights added here
until test block *just*
begins to slide over
test surface

▲ Figure 2.9 This apparatus can be used to investigate friction.

> A 'catch box' filled with bubble wrap (or similar) under the suspended masses keeps hands and feet out of the 'drop zone'.

The simple apparatus shown in Figure 2.9 can be used to discover some basic facts about friction. The weight force on the line running over the pulley pulls the block horizontally along the track and friction acts on the block to oppose this force. The weight is increased until the block just starts to move; this happens when the pull of the weight force just overcomes the friction force. The friction force between the block and the track has maximum value.

The apparatus can be used to test different factors that may affect the size of the friction force, such as the surfaces in contact – the bottom of the block and the surface of the track. If the track surface is replaced with a rough surface, like a sheet of sandpaper, the force required to overcome friction will be greater.

It is important to remember friction when you are investigating forces and motion. Friction affects almost every form of motion on Earth. However it is possible to do experiments in the science laboratory in which the friction force on a moving object is reduced to a very low value. Such an object can be set in motion with a small push and it will continue to move at a constant speed even when the force is no longer acting on it. An experiment like this is shown in Figure 2.10.

You may also have seen scientists working in space demonstrating that objects keep moving in a straight line at constant speed, once set in motion. They do this in space because the objects are **weightless** and the force of air resistance acting on them is very small.

EXTENSION WORK

Objects in orbit, such as spacecraft, are described as 'weightless' because they do not appear to have weight. However the Earth's gravity is still acting on them, and on the spacecraft. You can think of a spacecraft in orbit as 'falling around the Earth'. As the objects inside the spacecraft are also falling around the Earth at the same rate, they do not seem to fall inside the spacecraft.

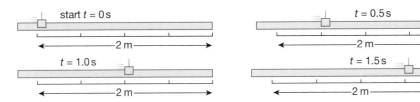

start $t = 0\,s$ $t = 0.5\,s$
2 m   2 m
$t = 1.0\,s$ $t = 1.5\,s$
2 m   2 m

▲ Figure 2.10 A linear (straight line) air-track reduces friction dramatically. The glider moves equal distances in equal time intervals. Its velocity is constant.

▲ Figure 2.11 Forces can cause changes in shape.

CHANGING SHAPES

We have seen that forces can make things start to move, accelerate or decelerate. The examples in Figure 2.11 show another effect that forces can have – they can change the shape of an object.

Sometimes the change of shape is temporary, as in the suspension spring in the mountain bike (Figure 2.11a). Sometimes the shape of the object is permanently changed, like a crushed can or a car that has collided with another object. A temporary change of shape may provide a useful way of absorbing and storing energy, as in the spring in a clock (Figure 2.11b). A permanent change may mean the failure of a structure like a bridge to support a load. Next we will look at temporary changes in the lengths of springs and elastic bands.

TEMPORARY CHANGES OF SHAPE

If you apply a force to an elastic band, its shape changes – the band stretches and gets longer. All materials will stretch a little when you put them under tension (that is, pull them) or shorten when you compress or squash them. You can stretch a rubber band quite easily, but a huge force is needed to cause a noticeable extension in a piece of steel of the same length.

Some materials, like glass, do not change shape easily and are brittle, breaking rather than stretching noticeably. Elastic materials do not break easily and tend to return to their original shape when the forces acting on them are removed, like the spring in Figure 2.11b. Other materials, like putty and modelling clay, are not elastic but plastic, and they change shape when even quite small forces are applied to them.

We will look at elastic materials, like rubber, metal wires and metals formed into springs, in the next part of this chapter.

SPRINGS AND WIRES

Springs are coiled lengths of certain types of metal, which can be stretched or compressed by applying a force to them. They are used in many different situations. Sometimes they are used to absorb raised bumps in the road as suspension springs in a car or bicycle. In beds and chairs they are used to make sleeping and sitting more comfortable. They are also used in door locks to hold them closed and to make doors close automatically.

To choose the right spring for a particular use, we must understand some important features of springs. A simple experiment with springs shows us that:

Springs change length when a force acts on them and they return to their original length when the force is removed.

This is true provided you do not stretch them too much. If springs are stretched beyond a certain point they do not spring back to their original length.

HOOKE'S LAW

Robert Hooke discovered another important property of springs. He used simple apparatus like that shown in Figure 2.12.

Hooke measured the increase in length (extension) produced by different load forces on springs. The graph he obtained by plotting force against extension was a straight line passing through the origin. This shows that the extension of the spring is proportional to the force. This relationship is known as Hooke's law.

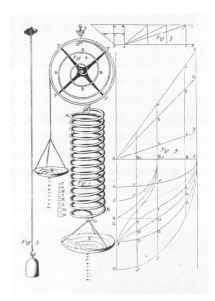

▲ Figure 2.12 Robert Hooke (1635–1703) was a contemporary of Sir Isaac Newton. This is a drawing of the apparatus Hooke used in his experimental work on the extension of a spring.

> ⚠ Wear eye protection if heavy loads are being used and clamp stands securely to a bench or table. Keep hands and feet away from beneath the load. Do not use excessively large loads.

ACTIVITY 1

▼ PRACTICAL: INVESTIGATE HOW THE EXTENSION OF A SPRING CHANGES WITH LOAD

Figure 2.13a shows apparatus that may be used to investigate the relationship between the force applied to a spring and its extension.

The length of the unstretched spring is measured with a half-metre rule then the spring is loaded with different weights. The extension for each load is measured against a scale using a set square to improve measurement accuracy.

A table of results is recorded and a graph of load force against extension plotted as shown in Figure 2.13b. The extension measurements can be checked by unloading the weights one at a time and remeasuring the extension for each load.

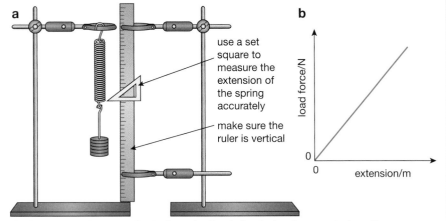

▲ Figure 2.13 **a** Apparatus for the investigation **b** Graph showing expected result

Here the result is a straight line graph passing through the origin of the axes. The spring obeys Hooke's law.

Hooke's law only applies if you do not stretch a spring too far. Figure 2.14 shows what happens if you stretch a spring too far. You can see that the line starts to curve at a point called the limit of proportionality. This is the point where the spring stops obeying Hooke's law and starts to stretch more for each increase in the load force. If the load is increased more, a point called the elastic limit is reached. Once you have stretched a spring beyond the elastic limit it will not return to its original length as you take the weights off the spring.

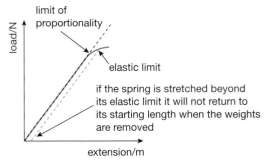

▲ Figure 2.14 Graph of load v extension showing spring going beyond elastic limit

Hooke's law also applies to wires. If you stretch a wire, you will find that the extension is proportional to the load up to a certain load then it may behave as the spring shown in Figure 2.14. Wires made of different metals will behave in different ways – some will obey Hooke's law until the wire breaks; other types of metal will stretch elastically and then plastically before breaking.

ELASTIC BANDS

> ⚠ Wear eye protection because if the elastic band breaks it can fly back with enough energy to cause serious eye damage.

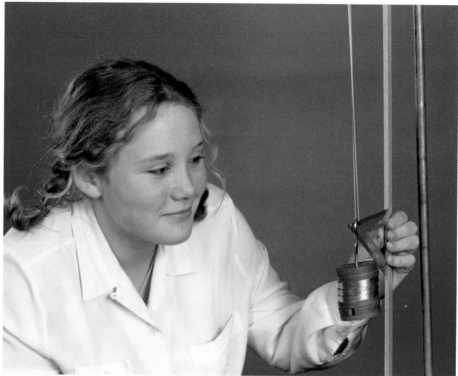

▲ Figure 2.15 Stretching an elastic band

Elastic bands are usually made of rubber. You can use the same apparatus shown in Figure 2.13a to investigate how an elastic band stretches under load. If you stretch an elastic band with increasing load forces, you get a graph like that shown in Figure 2.16. The graph is not a straight line, showing that elastic bands do not obey Hooke's law. You may also find that the extension produced by a given load force is different when you are increasing the load force to when you decrease the load force.

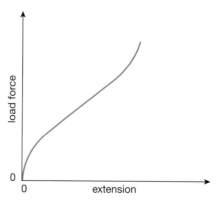

▲ Figure 2.16 Rubber bands do not obey Hooke's law – the extension is not directly proportional to the force causing it.

CHAPTER QUESTIONS

More questions on forces can be found at the end of Unit 1 on page 55.

SKILLS CRITICAL THINKING

1 Name the force that:
  a causes objects to fall towards the Earth
  b makes a ball rolled across level ground eventually stop
  c stops a car sinking into the road surface.

2 Name two types of force that oppose motion.

SKILLS PROBLEM SOLVING

3 The drawing shows two tug-of-war teams. Each person in the red team is pulling with a force of 250 N. Each person in the blue team is pulling with a force of 200 N.

  a What is the total force exerted by the blue team?
  b What is the total force exerted by the red team?
  c What is the resultant (unbalanced) force on the rope?
  d Which team will win?

SKILLS INTERPRETATION

4 The diagram below shows a block of wood on a sloping surface. Copy the diagram and label all the forces that act on the block of wood.

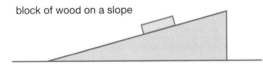

block of wood on a slope

5 A car is travelling along a level road at constant velocity (that is, its speed and direction are not changing). Draw a labelled diagram to show the forces that act on the car.

SKILLS CRITICAL THINKING

6 Describe two things that it would be impossible to do without friction.

SKILLS INTERPRETATION

7 Copy the diagrams below and label the direction in which friction is acting on the objects.

 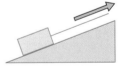

a a book on a sloping surface   **b** a block of wood being pulled up a slope

8 The drawing below shows a car pulling a caravan. They are travelling at constant velocity.

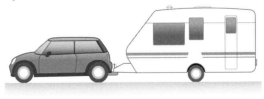

  a Copy the drawing. Label the forces that are acting on the caravan with arrows showing the direction that they act in.
  b Label the forces that act on the car.

9 The information in the following table was obtained from an experiment with a spring of original (unstretched) length 5 cm.

Load force on spring /newtons	Length of spring /cm	Extension of spring /cm
0	5.0	
2	5.8	
4	6.5	
6	7.4	
8	8.3	
10	9.7	
12	12.9	

a Copy and complete the table by calculating the extensions produced by each load.

b Use your table to plot a graph of force (y-axis) against extension (x-axis).

c Mark on your graph the part that shows Hooke's law.

d Sketch the shape of the graph you would expect if you carried out the same experiment using an elastic band instead of a spring.

3 FORCES AND MOVEMENT

The way an object moves depends upon its mass and the unbalanced force acting upon it. In this chapter you will find out how forces affect the way an object will move, particularly in the context of car safety.

a

b

▲ Figure 3.1 **a** This aircraft has only a short distance to travel before taking off and **b** a very short distance to land back on the aircraft carrier.

LEARNING OBJECTIVES

- Know and use the relationship between unbalanced force, mass and acceleration:

  force = mass × acceleration

  $F = m \times a$

- Know and use the relationship between weight, mass and gravitational field strength:

  weight = mass × gravitational field strength

  $W = m \times g$

- Know that the stopping distance of a vehicle is made up of the sum of the thinking distance and the braking distance

- Describe the factors affecting vehicle stopping distance, including speed, mass, road condition and reaction time

- Describe the forces acting on falling objects (and explain why falling objects reach a terminal velocity)

The aircraft in Figure 3.1 must accelerate to a very high speed in a very short time when taking off and decelerate quickly when landing back on the aircraft carrier. The unbalanced force on the plane causes the acceleration. The forces that act horizontally on the aircraft are the friction force between the wheels and the flight deck (where planes land on a ship), and air resistance, when the aircraft starts to move. At the start, the forward thrust of the aircraft engines is much greater than air resistance and friction, so there is a large unbalanced force to cause the acceleration. When the aircraft lands on the flight deck it must decelerate to stop in a short distance. Parachutes and drag wires are used to provide a large unbalanced force acting in the opposite direction to the aircraft's movement. An unbalanced force is sometimes referred to as a resultant force. In this chapter we look at how acceleration is related to the force acting on an object.

FORCE, MASS AND ACCELERATION

An object will not change its velocity (accelerate) unless there is an unbalanced force acting on it. For example, a car travelling along a motorway at a constant speed is being pushed along by a force from its engine, but this force is needed to balance the forces of friction and air resistance acting on the car. At a constant speed, the unbalanced force on the car is zero.

If there are unbalanced forces acting on an object, the object may accelerate or decelerate depending on the direction of the unbalanced force. The acceleration depends on the size of the unbalanced force and the mass of the object.

a When the same force is applied to objects with different mass, the smaller mass will experience a greater acceleration.

b Different-sized forces are applied to objects with the same mass. The small force produces a smaller acceleration than the large force.

▲ Figure 3.2 The acceleration of an object is affected by both its mass and the force applied to it.

INVESTIGATING FORCE, MASS AND ACCELERATION

Heavy wooden runways need to be stacked and moved carefully. They are best used at low level rather than being placed on benches or tables where they may fall off. If heavy trolleys are used as 'vehicles', a 'catch box' filled with bubble wrap or similar material should be placed at the end of the runway.

The experiment shown in Figure 3.3 shows how the relationship between force, mass and acceleration can be investigated. It uses a trolley on a slightly sloping ramp. The slope of the ramp is adjusted so that the trolley keeps moving down it if you push it gently. The slope is intended to overcome the friction in the trolley's wheels that could affect the results.

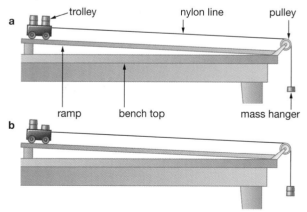

▲ Figure 3.3 You can use a trolley to find the acceleration caused by a particular force.

The force acting on the trolley is provided by the masses on the end of the nylon line. These masses accelerate as well as the trolley, so the force is increased by transferring one of the masses from the trolley to the mass hanger (Figure 3.3b). This increases the pulling force (explained later in this chapter) on the trolley, while keeping the total mass of the system the same.

The acceleration of the trolley can be measured by taking a series of pictures at equal time intervals using a digital video camera. Alternatively, a pair of light gates and a data logger can be used to find the speed of the trolley near the top of the ramp and near the end. The equation on page 9 can then be used to work out the acceleration.

EXTENSION WORK

The acceleration can be found using a digital video camera by measuring the distance travelled from the start for each image. Since the time between each image is known, a graph of displacement against time can be drawn. The gradient of the displacement–time graph gives the velocity at a particular instant, so data for a velocity–time graph can be obtained. The gradient of the velocity–time graph produced is the acceleration of the trolley.

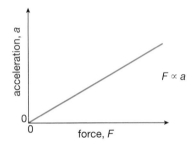

▲ Figure 3.4 Force is proportional to acceleration.

Figure 3.4 shows a graph of force against acceleration when the mass of the trolley and hanging masses is constant and the accelerating force is varied.

The graph is a straight line passing through the origin, which shows that:

force is proportional to acceleration

$$F \propto a$$

So doubling the force acting on an object doubles its acceleration.

In a second experiment, the accelerating force is kept constant and the mass of the trolley is varied.

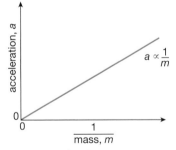

▲ Figure 3.5 Acceleration is inversely proportional to mass.

Figure 3.5 shows the acceleration of the trolley plotted against $\frac{1}{m}$. This is also a straight line passing through the origin, showing that:

acceleration is **inversely proportional** to mass

$$a \propto \frac{1}{m}$$

This means that for a given unbalanced force acting on an object, doubling the mass of the object will halve the acceleration.

Combining these results gives us:

force, F (N) = mass, m (kg) × acceleration, a (m/s^2)

$$F = m \times a$$

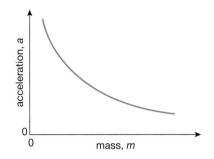

▲ Figure 3.6 The graph of a against m is a curve. Plotting a against $\frac{1}{m}$ (while keeping the force constant) gives a straight line. This makes it easier to spot the way that a is affected by m.

Force is measured in newtons (N), mass is measured in kilograms (kg), and acceleration is measured in metres per second squared (m/s²). From this we see that:

One newton is the force needed to make a mass of one kilogram accelerate at one metre per second squared.

▲ Figure 3.7 The equation can be rearranged using the triangle method. If you need to find the acceleration that results when a known force acts on an object of known mass, cover up a; you can see that $a = \frac{F}{m}$.

HINT

If an examination question asks you to write out the equation for calculating force, mass or acceleration, always give the actual equation (such as $F = m \times a$). You may not get the mark if you just draw the triangle.

DECELERATION IN A COLLISION

REMINDER

v is the final velocity, u is the initial velocity and t is the time for the change in velocity to take place.

REMINDER

The minus sign in Example 1 for velocity change indicates that the velocity has decreased.

If you are designing a car for high acceleration, the equation $F = ma$ tells you that the car should have low mass and the engine must provide a high accelerating force. You must also consider the force needed to stop the car.

When a moving object is stopped, it decelerates.

A negative acceleration is a deceleration.

If a large deceleration is needed then the force causing the deceleration must be large, too. Usually a car is stopped by using the brakes in a controlled way so that the deceleration is not excessive (too much). In an accident the car may collide with another vehicle or obstacle, causing a very rapid deceleration.

EXAMPLE 1

A car travelling at 20 m/s collides with a stationary lorry and stops completely in just 0.02 s. Calculate the deceleration of the car.

$$\text{acceleration} = \frac{\text{change in velocity}}{\text{time taken}} \text{ (see page 9)}$$

$$a = \frac{v - u}{t}$$

$$= \frac{0 \, \text{m/s} - 20 \, \text{m/s}}{0.02 \, \text{s}}$$

$$= \frac{-20 \, \text{m/s}}{0.02 \, \text{s}}$$

$$= -1000 \, \text{m/s}^2$$

A person of mass 50 kg in the car experiences the same deceleration when she comes into contact with a hard surface in the car. This could be the dashboard or the windscreen. Calculate the force that the person experiences.

$$F = m \times a$$

$$= 50 \, \text{kg} \times 1000 \, \text{m/s}^2$$

$$= 50\,000 \, \text{N}$$

In Chapter 4 you will learn about ways in which cars can be designed to reduce the forces on passengers in an accident.

FRICTION AND BRAKING

▲ Figure 3.8 Motorcycle disc brakes work using friction. Friction is necessary if we want things to stop.

The 'tread' of a tyre is the grooved pattern moulded into the rubber surface. It is designed to keep the rubber surface in contact with the road by throwing water away from the tyre surface.

Brakes on cars and bicycles work by increasing the friction between the rotating wheels and the body of the vehicle, as shown in Figure 3.8.

The friction force between the tyres and the road depends on the condition of the tyres and the surface of the road. It also depends on the weight of the vehicle. If the tyres have a good tread, are properly inflated (filled with air) and the road is dry, the friction force between the road and the tyres will be at its maximum.

Unfortunately, we do not always travel in ideal conditions. If the road is wet or the tyres are in bad condition the friction force will be smaller. If the brakes are applied too hard, the tyres will not grip the road surface and the car will skid (slide out of control). Once the car is skidding the driver no longer has control and it will take longer to stop. Skidding can be avoided by applying the brakes in the correct way, so that the wheels do not lock. Most modern cars are fitted with ABS (anti-lock braking system) to reduce the chance of a skid occurring. ABS is a computer-controlled system that senses when the car is about to skid and releases the brakes for a very short time.

SAFE STOPPING DISTANCE

The Highway Code used in the United Kingdom gives stopping distances for cars travelling at various speeds. The stopping distance is the sum of the thinking distance and the braking distance. The faster the car is travelling the greater the stopping distance will be.

THINKING DISTANCE

When a driver suddenly sees an object blocking the way ahead, it takes time for him or her to respond to the new situation before taking any action, such as braking. This time is called reaction time and will depend on the person driving the car. It will also depend on a number of other factors including whether the driver is tired or under the influence of alcohol or other drugs that slow reaction times. Poor visibility (for example, fog) may also make it difficult for a driver to identify a danger and so cause him or her to take longer to respond. Clearly, the longer the driver takes to react, the further the car will travel before braking even starts – that is, the longer the thinking distance will be. Equally clear is the fact that the higher the car's speed, the further the car will travel during this 'thinking time'. If the distance between two cars is not at least the thinking distance then, in the event of an emergency stop by the vehicle in front, a violent accident is inevitable.

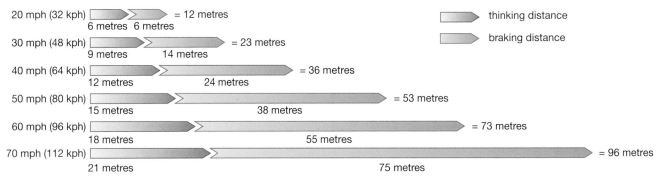

▲ Figure 3.9 The stopping distance is the distance the car covers from the moment the driver is aware of the need to stop to the point at which the vehicle comes to a complete stop.

BRAKING DISTANCE

With ABS (anti-lock braking system) braking, in an emergency you brake as hard as you can. This means that the braking force will be a maximum and we can work out the deceleration using the equation below.

$F = m \times a$, rearranged to give:

$a = \dfrac{F}{m}$ (shown in Figure 3.7)

It is worth pointing out here that vehicles with large masses, like lorries, will have smaller rates of deceleration for a given braking force – they will, therefore, travel further while braking.

Chapter 1 shows that the distance travelled by a moving object can be found from its velocity–time graph. The area under the graph gives the distance travelled. Look at the velocity–time graphs in Figures 3.10 and 3.11.

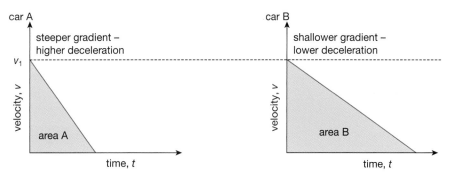

▲ Figure 3.10 Velocity–time graphs for two cars braking at different rates from the same speed, v_1, to rest.

Figure 3.10 shows two cars, A and B, braking from the same velocity. Car A is braking harder than car B and comes to rest in a shorter time. Car B travels further before stopping, as you can see from the larger area under the graph. Remember that the maximum rate of deceleration depends on how hard you can brake without skidding – in poor conditions the braking force will be lower.

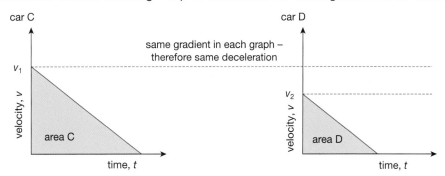

▲ Figure 3.11 Velocity–time graphs for two cars braking at the same rate to rest, from different speeds, v_1 and v_2.

Figure 3.11 shows two cars, C and D, braking at the same rate, as you can see from the gradients. Car C is braking from a higher velocity and so takes longer to stop. Again, the greater area under the graph for car C shows that it travels further whilst stopping than car D.

Vehicles cannot stop instantly! Remember also that the chart in Figure 3.9 shows stopping distances in ideal conditions. If the car tyres or brakes are in poor condition, or if the road surface is wet, icy or slippery, then the car will travel further before stopping.

WEIGHT

The weight of an object is the force that acts on it because of gravity. The weight of an object depends on its mass and the strength of gravity. The gravitational field strength (g) is the force that acts on each kilogram of mass. We can work out the weight of an object by using this equation:

weight, W (N) = mass, m (kg) × gravitational field strength, g (N/kg)

$$W = m \times g$$

You can use the triangle method to rearrange this equation.

▲ Figure 3.12 Rearranging $W = m \times g$ using the triangle method

Near and on the Earth's surface the gravitational field strength is approximately 9.8 N/kg, but we often use 10 N/kg to make calculations easier. The gravitational field strength on the Moon is about 1.6 N/kg, so an object taken from the Earth to the Moon will have less weight even though it has the same mass.

▲ Figure 3.13 An astronaut jumping on the Moon enjoying the effect of low gravity

EXAMPLE 2

An astronaut in a space suit with a complete life support pack has a mass of 140 kg. How much will the astronaut weigh a on the Earth, and b on Mars where the gravitational field strength is about one third of that on Earth? (Take the strength of the Earth's gravitational field as 10 N/kg.)

The force of gravity or weight of an object is given by:

weight, W = mass, m × gravitational field strength, g

a   weight on Earth = 140 kg × 10 N/kg

= 1400 N

b g on Mars = $\dfrac{10\,\text{N/kg}}{3}$ = 3.34 N/kg

weight on Mars = 140 kg × 3.34 N/kg

= 468 N

AIR RESISTANCE AND TERMINAL VELOCITY

An object moving through air experiences a force that opposes its movement. This force is called air resistance or drag. The size of the drag force acting on an object depends on its shape and its speed. Cars are designed to have a low **drag coefficient**. The drag coefficient is a measure of how easily an object moves through the air. High-speed trains have an efficient streamlined shape so that air flows more smoothly around them. Streamlined, smooth surfaces produce less drag.

It is particularly important to make fast-moving objects streamlined because the drag force increases with the speed of the object. The fact that drag increases with speed affects the way that dropped objects accelerate, because the faster they move the greater the drag force opposing their motion becomes.

Objects falling through the air experience two significant forces: the weight force (that is, the pull of gravity on the object) and the opposing drag force.

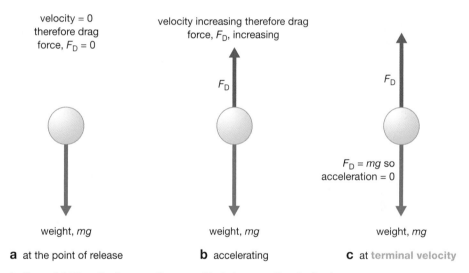

velocity = 0 therefore drag force, $F_D = 0$

velocity increasing therefore drag force, F_D, increasing

F_D

F_D

$F_D = mg$ so acceleration = 0

weight, mg weight, mg weight, mg

a at the point of release   **b** accelerating   **c** at **terminal velocity**

▲ Figure 3.14 How the forces acting on a object change as its velocity changes

In Figure 3.14a the object has just been released and has a starting velocity of 0 m/s. This means that there is no drag. (Remember that the drag force acts on moving objects.) The resulting downward-acting force is just the weight force. This force makes the object accelerate towards the Earth.

Figure 3.14b shows the object now moving. Because it is moving it has a drag force, F_D, acting on it. The drag force acts upwards (up) against the movement. This means that the resulting downward force (acting down) on the object is $(mg - F_D)$. You can see that the drag force has made the resulting downward force smaller, so the acceleration is smaller. All the time that the object is accelerating it is getting faster. The faster the object moves the bigger the drag force is.

In Figure 3.14c the drag force has increased to the point where it exactly balances the weight force – since there is now no unbalanced force on the object its acceleration is also zero. The object has reached its terminal velocity and although it is still falling it will not get any faster. Figure 3.15 shows a velocity–time graph for an object falling through air and reaching terminal velocity.

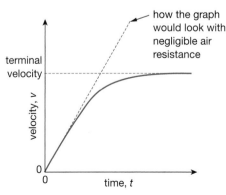

how the graph would look with negligible air resistance

terminal velocity

velocity, v

time, t

▲ Figure 3.15 The velocity–time graph for an object accelerating until it reaches terminal velocity

PARACHUTES

▲ Figure 3.16 These skydivers have just left the aeroplane. They will accelerate until they reach terminal velocity.

When a skydiver jumps from a plane she will accelerate for a time and eventually reach terminal velocity. Typically this will be between 150 and 200 kph. When she opens her parachute this will cause a sudden increase in the drag force. At this velocity (around 200 kph) the drag force of the parachute is greater than the weight of the skydiver. This means that the unbalanced force acting on the parachutist acts upwards and, for a while, she will decelerate. As she slows down the size of the drag force decreases and, eventually, a new terminal velocity is reached. Obviously the new terminal velocity depends on the design of the parachute, but it must be slow enough to allow the parachutist to land safely. Figure 3.17 shows a velocity–time graph for a skydiver.

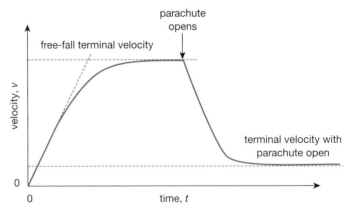

▲ Figure 3.17 Velocity–time graph for a free-fall parachutist reaching terminal velocity, then opening the parachute

MODELLING TERMINAL VELOCITY

Objects have to accelerate to quite high speeds in air to reach terminal velocity. This makes demonstrating the effect in a laboratory difficult. However, objects falling through liquids also experience a drag force that increases with speed. The sizes of drag forces in liquids are much higher than in gases. This means that objects falling through liquids have a much lower terminal velocity than objects falling through air, and can be used to model terminal velocity. You can use a tall measuring cylinder filled with water and drop small-**diameter** (1–2 mm) glass balls into it. Alternatively, use a much thicker liquid like oil and use small-diameter balls. As well as demonstrating terminal velocity this presents plenty of opportunities for investigations. You could measure the terminal velocity using a light gate and a data logger.

LOOKING AHEAD – WEIGHTLESSNESS

▲ Figure 3.18 Scientists in the International Space Station experience weightlessness

▲ Figure 3.19 Astronauts training for weightlessness in an aircraft

Figures 3.18 and 3.19 show astronauts in 'weightless' conditions in the International Space Station and training for weightlessness in an aircraft. In both cases the astronauts are still being pulled towards the Earth by gravity. They all experience the force we call weight, calculated using the equation *mg* that we met earlier in Chapter 3, yet they appear to float freely!

They are in a condition called free-fall. You may have noticed feeling very slightly 'lighter' when you are in a lift as it accelerates downwards. This effect is more noticeable if the lift accelerates at a greater rate. Further study of physics will explain all this.

Exploring the Solar System

Sir Isaac Newton worked out his laws of motion more than 300 years ago. Although Albert Einstein explained that these laws do not hold when objects are travelling at speeds close to the speed of light (3×10^8 m/s), Newton's laws give answers to calculations accurate enough for scientists to plan successful space missions like the Moon landing in 1969, the Phoenix lander to Mars in 2007 and the incredible achievement of landing the Philae lander (carried by the Rosetta spacecraft) on a comet in 2014.

If you look at Example 1 in Chapter 4 (based on the Saturn V rockets used in the Moon missions) the answer obtained using equation

$$\text{force} = \frac{\text{increase in momentum}}{\text{time taken}}$$ is not very accurate. There are a number of reasons for this: one is to do with air resistance, mentioned earlier in Chapter 3; another is to do with what is happening to the rocket as it uses up its fuel.

Understanding these allows physicists to apply the laws of motion more accurately.

CHAPTER QUESTIONS

More questions on force and acceleration can be found at the end of Unit 1 on page 55.

SKILLS CRITICAL THINKING

1 Explain what is meant by an unbalanced force? Illustrate your answer with an example.

SKILLS REASONING

2 Rockets burn fuel to give them the thrust needed to accelerate. As the fuel burns the mass of the rocket gets smaller. Assuming that the rocket motors provide a constant thrust force, explain what will happen to the acceleration of the rocket as it burns its fuel.

SKILLS PROBLEM SOLVING

3 a Calculate the force required to make an object of mass 500 g accelerate at 4 m/s^2. (Take care with the units!)

b An object accelerates at 0.8 m/s^2 when a resultant force of 200 N acts upon it. Calculate the mass of the object.

c What acceleration is produced by a force of 250 N acting on a mass of 25 kg?

SKILLS CRITICAL THINKING

4 Explain the meaning of the following terms used to describe stopping vehicles in an emergency:

a thinking distance

b braking distance

c overall stopping distance.

5 What factors affect the braking distance of a vehicle?

SKILLS ANALYSIS

6 The diagram below shows the velocity–time graph for a car travelling from the moment that the driver sees an object blocking the road ahead.

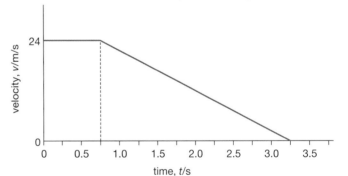

Use the graph to find out:

a how long the driver takes to react to seeing the obstacle (reaction time)

b how far the car travels in this reaction time

c how long it takes to bring the car to a halt once the driver starts braking

d the total distance the car travels before stopping.

SKILLS PROBLEM SOLVING

7 Calculate the weight of an apple of mass 100 grams:

a on the Earth (g = 10 N/kg)

b on the Moon (g = 1.6 N/kg).

SKILLS CRITICAL THINKING

8 What factors affect the drag force that acts on a high-speed train?

9 Describe an experiment to demonstrate terminal velocity. Say what measurements you need to take in your experiment to show that a falling object has reached terminal speed.

10 Look at the velocity–time graph for a free-fall parachutist shown below.

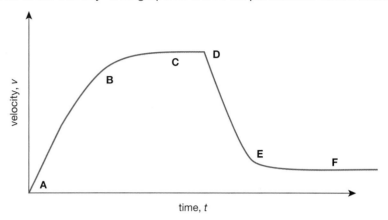

State and explain the direction and relative size of the unbalanced force acting on the parachutist at each of the points A–F labelled on the graph.

4 MOMENTUM

Momentum is possessed by masses in motion – it is calculated by multiplying the mass of an object by its velocity. In this chapter you will learn that, when objects speed up or slow down, the rate of change of momentum is proportional to the force causing the change. You will also see that momentum is conserved in collisions and explosions.

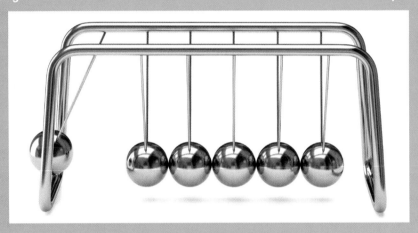

▲ Figure 4.1 Newton's cradle

LEARNING OBJECTIVES

■ Know and use the relationship between momentum, mass and velocity:

momentum = mass × velocity

$p = m \times v$

■ Use the idea of momentum to explain safety features

■ Use the conservation of momentum to calculate the mass, velocity or momentum of objects

■ Use the relationship between force, change in momentum and time taken:

$$force = \frac{change\ in\ momentum}{time\ taken}$$

$$F = \frac{(mv - mu)}{t}$$

■ Demonstrate an understanding of Newton's third law

KEY POINT

Do not confuse momentum, a property of moving masses, with moment, the turning effect of a force.

Newton's cradle is an entertaining toy but it also demonstrates a physics conservation law. When one of the balls is drawn back a short way and released, it swings and collides with the remaining group of balls. After the impact the ball at the opposite end springs away and swings out as far as the first ball was drawn back to start with. If two balls are drawn back and released then two balls will move away at the opposite end as the collision occurs. The moving ball has momentum, and momentum is conserved in collisions. This chapter is about momentum and what is meant by conservation of momentum.

MOMENTUM

We talk about objects 'gaining momentum' in everyday speech. When we say this we are usually trying to get across an idea of something becoming more difficult to stop. Sometimes we use the word in a way that is very close to the way it is defined in physics – for example, if a car starts rolling down a hill we might say that the car is 'gaining momentum' as it speeds up.

In physics, momentum is a quantity possessed by masses in motion. Momentum is a measure of how difficult it is to stop something that is moving. We calculate the momentum of a moving object using the equation:

momentum, p (kg m/s) = mass, m (kg) × velocity, v (m/s)

$$p = m \times v$$

Momentum is a vector quantity and is measured in kilogram metres per second (kg m/s) – provided that mass in the above equation is measured in kg and velocity in m/s.

You can see from the equation that the more mass an object has, the more momentum it will have when moving. The momentum of a moving object also increases with its speed.

MOMENTUM AND ACCELERATION

We have already discussed the relationship $F = m \times a$ (see page 30). This relationship was first discussed by Sir Isaac Newton (1642–1727). A more precise statement of Newton's discovery would be that when an unbalanced force acts on an object it causes a change in the momentum of the object in the direction of the unbalanced force. Newton discovered that the rate of change of momentum of an object is proportional to the force applied to that object. This means that if you double the force acting on an object its momentum will change twice as quickly. Figure 4.2 shows the effect of the thrust force on the velocity of a space shuttle, and, therefore, on its momentum.

a

initial velocity, u m/s

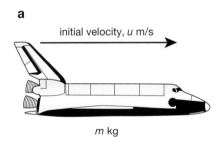

m kg

b

force, F, causing acceleration

F N

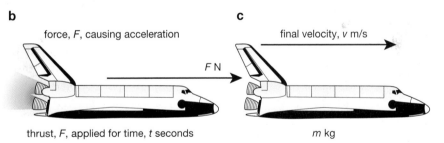

thrust, F, applied for time, t seconds

c

final velocity, v m/s

m kg

▲ Figure 4.2 The thrust of the rocket motor makes the velocity and, therefore, the momentum of the shuttle increase.

initial momentum of object = mu

final momentum of object = mv

therefore, increase in momentum = $mv - mu$

so, rate of increase of momentum = $\dfrac{(mv - mu)}{t}$

As stated above, Newton identified a proportional relationship between the rate of increase of momentum and the force applied, but with the system of units we use the relationship appears as shown:

$$\text{force, } F = \frac{\text{change in momentum } (mv - mu)}{\text{time taken, } t}$$

$$F = \frac{mv - mu}{t}$$

If you look at this equation you will notice it can be rearranged to give the more familiar equation $F = ma$:

$$F = \frac{mv - mu}{t}$$

$$= \frac{m(v - u)}{t}$$

since $\quad \dfrac{(v - u)}{t} = a$

then $\quad\quad\quad F = ma$

The rearrangement is possible because we have assumed that the mass of the object involved is constant. This will not always be the case in real situations – for example, when a space shuttle is launched the mass of the rocket changes continuously as fuel is burned and rocket stages are jettisoned. Rocket scientists also have to include air resistance in their calculations. However, you will not meet problems like this at International GCSE level.

EXAMPLE 1

The first stage of the type of rocket used in Moon missions provides an unbalanced upward (away from the Earth) force of 30 MN and burns for 2.5 minutes.

a Calculate the increase in the rocket's momentum that results.

b If the rocket has a mass of 3000 tonnes what is the velocity of the rocket after the first stage has completed its 'burn'?

(Take care to convert units into N, kg and s. 1 MN = 10^6 N; 1 tonne = 1000 kg)

a The equation $F = \dfrac{(mv - mu)}{t}$ is another that can be rearranged using the triangle.

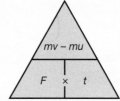

▲ Figure 4.3 Rearranging the change in momentum = $f \times t$ equation using the triangle method

We want the change in momentum (at the top of the triangle) so it is at the bottom of the triangle:

$$F \times t = (3 \times 10^7 \text{ N}) \times (2.5 \times 60 \text{ s})$$

increase in momentum is 4.5×10^9 kg m/s

b The rocket starts from rest, $u = 0$, therefore the initial momentum (mu) is 0. This means that the increase in momentum is $mv = 4.5 \times 10^9$ kg m/s. Divide this by mass, $m = 3 \times 10^6$ kg, to give:

$$v = \frac{4.5 \times 10^9 \text{ kg m/s}}{3 \times 10^6 \text{ kg}}$$

$$= 1.5 \times 10^3 \text{ m/s}$$

As we have seen above this has been calculated with the mass of the rocket staying the same. In a real launch fuel is used up so the mass of the rocket will decrease and the calculation would be harder.

MOMENTUM AND COLLISIONS

The total momentum of objects that collide remains the same:

momentum before the collision = momentum after the collision

You will need to use this to work out what happens to objects that collide when they are moving along the same straight line.

You will not need to learn the following explanation for the exam, but you do need to understand Newton's third law. You will see this again later in this chapter.

We can express the above equation in terms of momentum change, as follows:

force × time = increase in momentum

This simply says that a bigger force applied to an object for a longer time will result in a greater change in the momentum of the object.

Consider what happens when two balls collide, as shown in Figure 4.4.

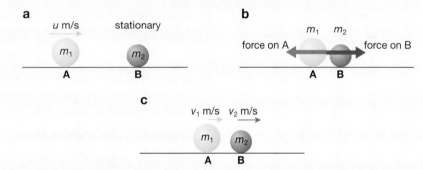

▲ Figure 4.4 **a** Moving ball A, mass m_1, rolls towards stationary ball B, mass m_2; **b** During the impact each ball exerts a force on the other – equal in size and opposite in direction; **c** The balls after the collision.

During the time the two balls are in contact each exerts a force on the other (Newton's third law about action and reaction, see page 46). The forces act in opposite directions and obviously act for the same amount of time. This means that $F \times t$ for each is the same size, but opposite in direction.

The term impulse is used for the product (force × time).

The increase in momentum of ball B is exactly balanced by the decrease in momentum of ball A, so the total momentum of the two balls is unchanged before and after the collision – momentum is conserved.

In any system, momentum is always conserved provided no external forces act on the system. This means that when two snooker balls collide the momentum of the balls is conserved if no friction forces act on them. The presence of friction means that the balls will eventually slow down and stop, thus ending up with no momentum. Although the balls have 'lost' momentum something else will, inevitably, have gained an equal amount of momentum! As the balls are slowed by the friction of the snooker table they, in turn, cause a friction force to act on the table. The table gains some momentum. However, the large mass of the table means that the effect is unnoticeable.

EXAMPLE 2

A railway truck with a mass of 5000 kg rolling at 3 m/s collides with a stationary truck of mass 10 000 kg. The trucks join together. At what speed do they move after the collision?

We shall assume that friction forces are small enough to ignore, so we can apply the law of conservation of momentum:

momentum before the collision = momentum after the collision

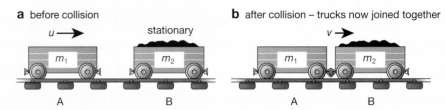

a before collision

$u \longrightarrow$ stationary

m_1 m_2

A   B

b after collision – trucks now joined together

$v \longrightarrow$

m_1 m_2

A   B

▲ Figure 4.5 Railway trucks in collision

so, momentum of A before collision + momentum of B before collision = momentum of A and B moving together after collision

$$(m_1 \times u) + (m_2 \times 0 \text{ m/s}) = (m_1 + m_2) \times v$$

where m_1 is the mass of truck A, u is its velocity before the collision, m_2 is the mass of truck B (at rest before the collision so its velocity is 0 m/s), and v is the velocity of the two trucks after the collision.

Substituting these values gives:

$$(5000 \text{ kg} \times 3 \text{ m/s}) + (10\,000 \text{ kg} \times 0 \text{ m/s}) = (5000 \text{ kg} + 10\,000 \text{ kg}) \times v$$

so $v = \dfrac{15\,000 \text{ kg m/s}}{15\,000 \text{ kg}} = 1 \text{ m/s}$

After the collision the trucks move with a velocity of 1 m/s in the same direction that the original truck was travelling.

EXPLOSIONS

▲ Figure 4.6 A shuttle launch

The conservation of momentum principle can be applied to explosions. An explosion involves a release of energy causing things to fly apart. The momentum before and after the explosion is unchanged, though there will be a huge increase in movement energy. A simple demonstration of a safe 'explosion' is shown in Figure 4.6. You might think that the rocket in this photo is taking off because the rocket motor is pushing against the ground, but rockets work in space where there is nothing to push against.

Rocket motors use the principle of conservation of momentum to propel spacecraft through space. They produce a continuous, controlled explosion that forces large amounts of fast-moving gases (produced by the fuel burning) out of the back of the rocket. The spacecraft gains an equal amount of momentum in the opposite direction to that of the moving exhaust gases.

You can see the same effect if you blow up a balloon and release it without tying up the end!

CAR SAFETY

In Example 1 on page 31 you saw that the force on a person in a car crash can be very large. In that example, the force was worked out from the deceleration in a crash. You can also work out the force in a crash using the equation for momentum.

Steep roads often have escape lanes filled with deep, soft sand. The soft sand slows heavy lorries that are out of control slowly – by making time, t, for the lorry to stop longer, the force, F, slowing the lorry is smaller and the driver is less likely to suffer serious injury,

EXAMPLE 3

A car travelling at 20 m/s collides with a stationary lorry and is brought to rest in just 0.02 s. A woman in the car has a mass of 50 kg. She experiences the same deceleration when she comes into contact with a hard surface in the car (such as the dashboard or the windscreen). What force does the person experience?

$$\text{force} = \frac{\text{change in momentum}}{\text{time}}$$

$$= \frac{(50\ \text{kg} \times 20\ \text{m/s}) - (50\ \text{kg} \times 0\ \text{m/s})}{0.02\ \text{s}}$$

$$= 50\,000\ \text{N}$$

▲ Figure 4.7 Parts of cars are designed to crumple.

Cars are now designed with various safety features that increase the time over which the car's momentum changes in an accident. Figure 4.7 shows the safety features of a car being tested. The car has a rigid passenger cell or compartment with crumple zones in front and behind. The crumple zones, as the name suggests, collapse during a collision and increase the time during which the car is decelerating. For instance, if the deceleration time in Example 3 above is increased from 0.02 s to 1 s, then the impact causes a much smaller force of just 1000 N to act on the passenger, greatly increasing their chances of survival.

Crumple zones are just one of the safety features now used in modern cars to protect the passengers in an accident. They only work if the passengers are wearing seat belts so that the reduced deceleration applies to their bodies too. Without seat belts, the passengers will continue moving forward until they come into contact with some part of the car or with a passenger in front. If they hit something that does not crumple they will be brought to rest in a very short time, which, as we have seen in Example 3, means a large deceleration and, therefore, a large force acting on them.

NEWTON'S LAWS OF MOTION

EXTENSION WORK

Scientists use the word 'law' very cautiously. Only when a hypothesis (idea) has been tested many times independently by careful experiment is it raised to the status of a 'law'. Einstein showed that in special situations Newton's laws break down, but they are still accurate enough to predict the way objects respond to forces with a high degree of accuracy.

EXTENSION WORK

Sir Isaac Newton lived from 1642 to 1727. He made many famous discoveries and some important observations about how forces affect the way objects move. The first observation, called Newton's first law, was:

> *Things don't speed up, slow down or change direction unless you push (or pull) them.*

Newton didn't put it quite like that, of course! He said that a body would continue to move in a straight line at a steady speed unless a resultant (unbalanced) force was acting on the body. If the forces acting on an object are balanced then a stationary object stays in one place, and a moving object continues to move in just the same way as it did before the forces were applied. You have already met this idea in Chapter 3.

Newton then asked another obvious question: how does the acceleration of an object depend on the force that you apply to it? Again Newton's formal statement of the answer (Newton's second law) sounds complicated, but the basic findings are quite simple:

> *The bigger the force acting on an object, the faster the object will speed up.*

Objects with greater mass require bigger forces than those with smaller mass to make them speed up (accelerate) at the same rate.

The force referred to is the resultant force acting on the object. This idea is also covered in Chapter 3. The relationship $F = m \times a$ is a consequence of Newton's second law.

The only law you you need to learn by name is Newton's third law.

NEWTON'S THIRD LAW: ACTION AND REACTION

In simple language Newton's third law might be stated:

> *When you push something it pushes you back just as hard, but in the opposite direction.*

It is usually stated as:

> *For every action there is an equal and opposite reaction.*

When you sit down, your weight pushes down on the seat. The seat pushes back on you with an equal, but upward, force. An experiment to give you the idea of what this law means is shown in Figure 4.8.

In Figure 4.8, person X is clearly pushing person Y but it is not obvious that Y is pushing X back. When both X and Y move it is clear that X has been affected by a force pushing him to the left. The force felt by X is the reaction force.

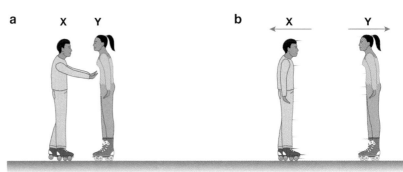

▲ Figure 4.8 For every action there is an equal and opposite reaction.

Look back at Figure 2.3 on page 19. The weight of the car is pushing down on the ground. Although the weight is shown coming from the centre of gravity of the car, it acts through the wheels pressing down on the ground. The total reaction force from the ground acting upwards on the wheels is the same as the weight of the car. It is this reaction force that stops the car sinking into the ground.

It can sometimes be difficult to sort out action and reaction forces from balanced forces. Balanced forces act in opposite directions on the same object. Action and reaction forces also act in opposite directions, but are always acting on different objects.

CHAPTER QUESTIONS

More questions on momentum can be found at the end of Unit 1 on page 55.

SKILLS   PROBLEM SOLVING

1 Work out, giving your answers in kg m/s, the momentum of the following moving objects:

    a   a bowling ball of mass 6 kg travelling at 8 m/s

    b   a ship of mass 50 000 kg travelling at 3 m/s

    c   a tennis ball of mass 60 g travelling at 180 km/h.

2 An air rifle pellet of mass 2 g is fired into a block of modelling clay mounted on a model railway truck. The truck and modelling clay have a mass of 0.1 kg. The truck moves off after the pellet hits the modelling clay with an initial velocity of 0.8 m/s.

    a   Calculate the momentum of the modelling clay and truck just after the collision.

    b   State the momentum of the pellet just before it hits the modelling clay.

    c   Use your answers to a and b to calculate the velocity of the pellet just before it hits the modelling clay.

SKILLS   REASONING

    d   State any assumptions you made in this calculation.

SKILLS   PROBLEM SOLVING

3 A rocket of mass 1200 kg is travelling at 2000 m/s. It fires its engine for 1 minute. The forward thrust provided by the rocket engines is 10 kN (10 000 N).

    a   Use increase in momentum = $F \times t$ to calculate the increase in momentum of the rocket.

    b   Use your answer to a to calculate the increase in velocity of the rocket and its new velocity after firing the engines.

SKILLS   CRITICAL THINKING

4 Air bags are a safety feature fitted to all modern cars. In the event of sudden braking an air bag is rapidly inflated with gas in front of the driver or passenger.

    a   Does the air bag remain inflated?

SKILLS   REASONING

    b   Explain how the air bag protects the driver from more serious injury.

END OF PHYSICS ONLY

5 THE TURNING EFFECT OF FORCES

A force can have a turning effect – it can make an object turn around a fixed pivot point. When the anticlockwise turning effects of forces are balanced by turning forces in the clockwise direction, the object will not turn – it is in balance.

▲ Figure 5.1 Turning effects are used in many places, such as in **a** the see-saw, and **b** the cranes.

LEARNING OBJECTIVES

- Know and use the relationship between the moment of a force and its perpendicular distance from the pivot:

  moment = force × perpendicular distance from the pivot

- Know that the weight of a body acts through its centre of gravity

- Use the principle of moments for a simple system of parallel forces acting in one plane

- Understand how the upward forces on a light beam, supported at its ends, vary with the position of a heavy object placed on the beam

Unbalanced forces acting on objects can make them accelerate or decelerate. In the examples in Figure 5.1, the forces acting are having a turning effect. They are making the objects, like the see-saw, turn around a fixed point called a **pivot** or fulcrum. The turning effect in cranes is explained in Figure 5.6 on page 51. We use this turning effect of forces all the time. In our bodies the forces of our muscles make parts of our bodies turn around joints like our elbows or knees. When you turn a door handle, open a door or remove the lid off a tin of paint with a knife you are using the turning effect of forces. Understanding the turning effect of forces is important. Sometimes we want things to turn or rotate – the see-saw wouldn't be much fun if it didn't. However, sometimes we want the turning effects to balance so that things don't turn – it would be terrible if the crane did not balance!

OPENING A DOOR

Challenge a partner (perhaps one who thinks that he or she is strong!) to try to hold a door closed while you try to open it – then explain the rules! They can apply a pushing force but no further than 20 cm from the hinge (the part of the door that fastens it to the wall), while you try to pull the door open by pulling on the handle. You will be able to open the door quite easily.

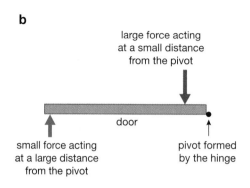

Figure 5.2 The distance of the force from the pivot is crucial.

You will realise that you have an advantage because the turning effect of your force doesn't just depend on the size of the force you apply but also on the distance from the hinge or pivot at which you apply it. You have the advantage of greater leverage (Figure 5.2).

THE MOMENT OF A FORCE

The turning effect of a force about a hinge or pivot is called its moment. The moment of a force is defined like this:

moment of a force (Nm) = force, F (N) × perpendicular distance from pivot, d (m)

$$moment = F \times d$$

The moment of a force is measured in newton metres (Nm) because force is measured in newtons and the distance to the pivot is in metres. We need to be precise about what distance we measure when calculating the moment of a force. Look at the diagrams in Figure 5.3.

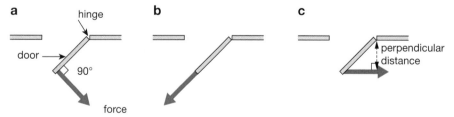

Figure 5.3 For a force to have its biggest effect it should be at 90° to the lever.

If you think about the simple door opening 'competition' we discussed earlier you will realise that for a force to have the biggest turning effect it should be applied as in Figure 5.3a – that is, its line of action should be perpendicular (at 90°) to the door.

In Figure 5.3b the force has no turning effect at all because the line along which the force is acting passes through the pivot.

Figure 5.3c shows how the distance to the pivot must be measured to get the correct value for the moment. The distance is the perpendicular distance from the line of action of the force to the pivot.

IN BALANCE

An object will be in balance (that is, it will not try to turn about a pivot point) if:

sum of anticlockwise moments = sum of clockwise moments

For example, Figure 5.4 shows two children sitting on a see-saw.

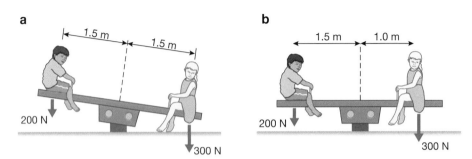

▲ Figure 5.4 The moment of the heavier child is reduced when she sits closer to the pivot, so that it balances the moment of the lighter child.

In Figure 5.4a:

the anticlockwise moment = 200 N × 1.5 m = 300 Nm
the clockwise moment = 300 N × 1.5 m = 450 Nm

So the see-saw is not balanced and leans down to the right as it rotates clockwise about the pivot.

The calculations for Figure 5.4a are approximate because the forces are not acting perpendicularly to the see-saw.

In Figure 5.4b:

the anticlockwise moment = 200 N × 1.5 m = 300 Nm

the clockwise moment = 300 N × 1.0 m = 300 Nm

So the see-saw is now balanced.

EXAMPLE 1

Look again at Figure 5.2. Person A pushes the door with a force of 200 N at a distance of 20 cm from the hinge (the pivot in this example). Person B opens the door by pulling the door handle which is 80 cm from the hinge.

What is the minimum pulling force that person B must use to open the door?

Here is a simplified sketch of the problem:

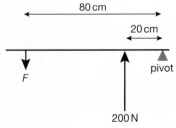

▲ Figure 5.5 A labelled sketch of the problem

The door will be in balance (on the point of opening) if:

sum of anticlockwise moments = sum of clockwise moments

$$F \times 80 \text{ cm} = 200 \text{ N} \times 20 \text{ cm}$$

$$F = \frac{200 \text{ N} \times 20 \text{ cm}}{80 \text{ cm}}$$

$$= 50 \text{ N}$$

But this means the turning effects are balanced and we want person B to apply a big enough force to make the anticlockwise moment bigger than the clockwise moment.

The answer is that person B must apply a force greater than 50 N ($F > 50$ N).

Look back at Figure 5.1 and look at Figure 5.6. The load arm is long so that the crane can reach across a construction site and move loads backwards and forwards along the length of the arm. The weight of the long load arm and the load must be counterbalanced by the large concrete blocks at the end of the short arm that projects out behind the crane controller's cabin. The counterbalance weights must be large because they are positioned closer to the pivot point, where the crane tower supports the crosspiece of the crane. Without careful balance the turning forces on the support tower could cause it to bend and collapse.

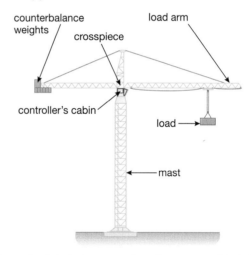

Figure 5.6 The clockwise and anticlockwise moments on the crane must be balanced.

CENTRE OF GRAVITY

Try balancing a ruler on your finger, as shown in Figure 5.7.

When the ruler is balanced, the anticlockwise moment is equal to the clockwise moment, but there are no downward forces acting in this situation other than the weight of the ruler itself. We know that the weight of the ruler is due to the pull of the Earth's gravity on the mass of the ruler. The mass of the ruler is equally spread throughout its length. It is not, therefore, surprising to find that the ruler balances at its centre point.

We say that the centre of gravity of the ruler is at this point. The weight force of the ruler acts through the centre of gravity – it is the point where the whole of the weight of the ruler appears to act. If we support the ruler at this point there is no turning moment in any direction about the point, and it balances. The centre of gravity is sometimes called the centre of mass.

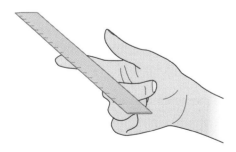

Figure 5.7 Can you find the point at which the ruler balances?

Finding the centre of gravity of a sheet of card

If you have a sheet of card or any other uniform material then finding its centre of gravity is quite straightforward – it will be located where the axes of symmetry cross, as shown in Figure 5.8.

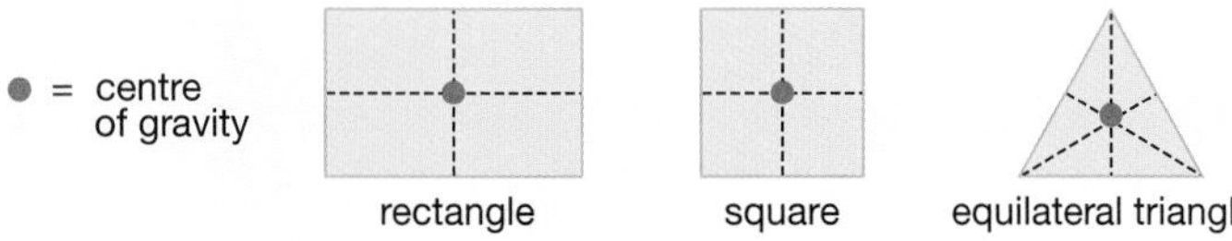

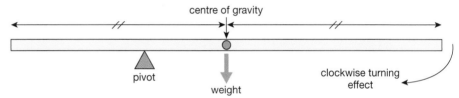

● = centre of gravity

rectangle   square   equilateral triangle

▲ Figure 5.8 The centre of gravity for these three regular shapes is where the axes of symmetry cross. An axis of symmetry can be found using a plane mirror. If you place a plane mirror along one of the dotted lines in any of the above shapes, the reflection in the mirror looks exactly like the original.

OBJECTS NOT PIVOTED AT THE CENTRE OF GRAVITY

A simple see-saw is a uniform beam (plank) pivoted in the middle. The centre of gravity of a uniform beam is in the middle, so the see-saw is pivoted through its centre of gravity. When an object is not pivoted through its centre of gravity the weight of the object will produce a turning effect. This is shown in Figure 5.9.

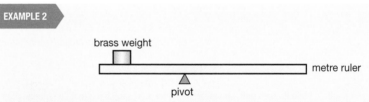

centre of gravity

pivot

weight

clockwise turning effect

▲ Figure 5.9 The weight of the beam causes a clockwise turning effect about the pivot.

brass weight

metre ruler

pivot

▲ Figure 5.10 In balance – not tipping in either direction

This sketch diagram shows a uniform wooden metre ruler with a mass of 0.12 kg. A pivot at the 40 cm mark supports the rule. When a brass weight is placed at the 20 cm mark the ruler just balances in the horizontal position. Take moments about the pivot to calculate the mass of the brass weight.

The weight of the ruler acts vertically through the centre of gravity of the rule. As the ruler is uniform the centre of gravity is at the 50 cm mark.

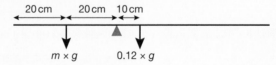

20 cm   20 cm   10 cm

$m \times g$ $0.12 \times g$

▲ Figure 5.11 A labelled sketch of the problem

For balance:

anticlockwise moments = clockwise moments

$$(m \times g) \times 20 \text{ cm} = (0.12 \text{ kg} \times g) \times 10 \text{ cm}$$

The mass of the brass weight is 0.06 kg.

Here g is the gravitational field strength in N/kg but, as you can see, it cancels. You can leave the distances in centimetres because you are using the same length units on both sides of the equation.

Centre of gravity and stability

The position of the centre of gravity of an object will affect its stability. A stable object is one that is difficult to push over – when pushed and then released it will tend to return to its original position.

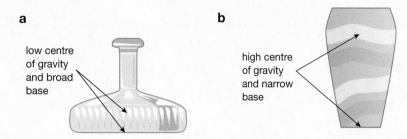

a   low centre of gravity and broad base

b   high centre of gravity and narrow base

▲ Figure 5.12 Stable and **unstable** objects

The shape in Figure 5.12a is typical for a ship's decanter (bottle). Bottles used on ships need to be difficult to knock over for obvious reasons! It is stable because, when knocked, its low centre of gravity and wide base result in a turning moment that tries to pull it back to its original position. The object in Figure 5.12b has a higher centre of gravity and smaller base, so it is much less stable. Only a small displacement is needed to make it fall over.

FORCES ON A BEAM

Figure 5.13a shows a boy standing on a beam across a stream. The beam is not moving, so the upward and downward forces must be balanced. As the boy is standing in the middle of the beam, the upward forces on the ends of the beam are the same as each other. The forces are shown in Figure 5.13b – to keep things simple, we are ignoring the weight of the beam.

If he moves to one end of the beam, as shown in Figure 5.13c, then the upward force will all be at that end of the beam. As he moves along the beam, the upward forces at the ends of the beam change. In Figure 5.13d, he is one-quarter along the beam. The upward force on the support nearest to him is three-quarters of his weight, and the upward force on the end furthest away from him is only one-quarter of his weight.

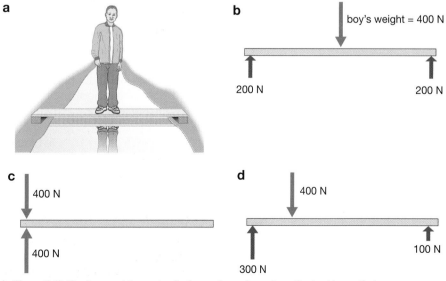

▲ Figure 5.13 The forces at the ends of a beam depend on where the load is applied.

CHAPTER QUESTIONS

More questions on the turning effect of forces can be found at the end of Unit 1 on page 55.

SKILLS ANALYSIS

1 Look at the diagram below. It shows various forces acting on objects about pivots.

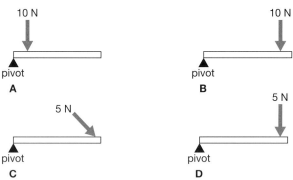

Write down the letter of the diagram that:

a shows the greatest turning moment about the pivot

b shows a turning moment of zero.

2 Look at the four see-saw diagrams below.

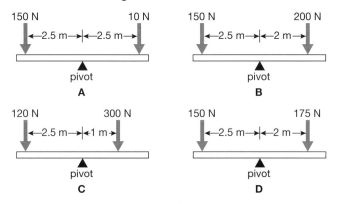

Write down the letter(s) of the diagram(s) that agree with the following statements, or write 'none' if no diagram agrees with the statement.

a The see-saw is balanced.

b The see-saw will rotate clockwise (dip down to the right).

c The see-saw will rotate anti-clockwise (dip down to the left).

3 A bookshelf is 2 m long, with supports at its ends (P and Q). Do not consider the weight of the bookshelf when you are answering these questions.

a Draw a sketch of the shelf, showing the supports.

SKILLS PROBLEM SOLVING

b A book weighing 10 N is placed in the middle of the shelf. What are the upward forces at P and Q?

c The book is moved so that it is 50 cm from Q. Use moments to calculate the forces at P and Q.

d The bookshelf weighs 10 N. Repeat parts b and c taking into account the weight of the shelf as well as the weight of the book.

UNIT QUESTIONS

SKILLS CRITICAL THINKING **1**

a Which of the following quantities is not a vector quantity?

A force

B area

C displacement

D velocity

(1)

b Which of the following quantities is not a scalar quantity?

A speed

B your age

C volume

D acceleration

(1)

c Which of the following statements about the motion of an object on which unbalanced forces act is false?

A The object could continue moving at constant speed in a straight line.

B The object could accelerate.

C The object could slow down.

D The object could change the direction in which it is moving.

(1)

d A ball is thrown vertically upwards in the air. Which of the following statements about the motion of the ball is false?

A The ball will be travelling as fast as it was when it gets back to the ground as it was when first thrown upwards.

B The ball will be stationary for an instant at the highest point in its flight.

C The direction of the force on the ball changes so the ball falls back to the ground.

D The direction of the acceleration is always downwards.

(1)

(Total for Question 1 = 4 marks)

SKILLS ANALYSIS

6

2 A tennis ball is dropped from a height of 3 m and bounces back to a height of 1 m after hitting the ground ($s = 0$).

a Which of the following distance–time graphs shows this motion correctly?

(1)

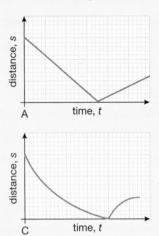

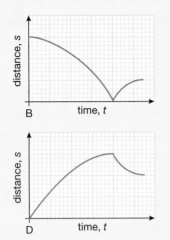

SKILLS CRITICAL THINKING

7

b i Explain the difference between the terms distance and displacement.

(2)

SKILLS PROBLEM SOLVING

8

ii What is the displacement of the tennis ball at the end of the described motion?

(1)

9

c i Use the equation $v^2 = u^2 + 2as$ to calculate the velocity of the tennis ball when it hits the ground.

(4)

7

ii What is the average speed of the tennis ball while it is falling to the ground?

(2)

8

iii Calculate how long it takes for the tennis ball to reach the ground.

(2)

SKILLS CRITICAL THINKING

6

d How can the distance travelled by the tennis ball be found from a velocity–time graph of its motion?

(1)

SKILLS INTERPRETATION

8

e Sketch a velocity–time graph for the tennis ball from the time that it is released to the time that it has rebounded to 1 m after the first bounce. Assume that the time that the ball is in contact with the ground is negligible (too short to show on your graph). Include as much numerical detail as you can without doing any further calculations.

(6)

(Total for Question 2 = 19 marks)

SKILLS CRITICAL THINKING

3 A student uses the following piece of apparatus to investigate how the extension of a length of wire varies with the force applied to it.

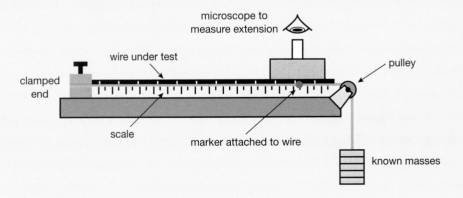

SKILLS DECISION MAKING

SKILLS EXECUTIVE FUNCTION

SKILLS CRITICAL THINKING

a   **i**   Name the independent variable in this investigation.   **(1)**

    ii   Name the dependent variable in this investigation.   **(1)**

    iii   Name one control variable in this investigation.   **(1)**

b   How is the force being applied to the wire calculated?   **(1)**

c   Explain how the student should use the apparatus to determine if the wire obeys Hooke's law.   **(5)**

d   The student also wants to know if the sample behaves elastically. Explain how the student can improve the investigation and discover whether or not the wire behaves elastically.   **(3)**

(Total for Question 3 = 12 marks)

PHYSICS ONLY

This question is about momentum.

a   State whether momentum is a scalar or a vector quantity and explain your answer.   **(3)**

The following apparatus is used to investigate the momentum of two gliders on a linear air-track.

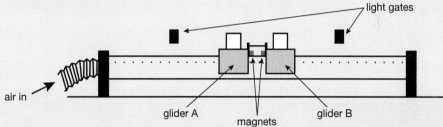

The gliders are fitted with magnets with like poles facing each other, so that they repel each other, and tied together with a cotton thread. This cotton thread will be burnt using a flame later in the investigation. 5 cm square cards are attached to both gliders – these will pass through the light gates. Digital timers time how long each card takes to pass through the light gates. The gliders are initially at rest.

SKILLS PROBLEM SOLVING

SKILLS INTERPRETATION

SKILLS CRITICAL THINKING

SKILLS REASONING

SKILLS PROBLEM SOLVING

b   State the initial momentum of the gliders.   **(1)**

c   **i**   Draw a labelled diagram showing the forces acting on glider A before the thread is burnt.   **(4)**

    ii   State Newton's third law and explain how it applies to the horizontal forces on the gliders when the thread is cut with the flame.   **(3)**

    iii   Describe what happens to the two gliders when the thread is cut.   **(1)**

d   After the thread is cut the card on glider A passes the light gate in 1.25 s.

    i   Calculate the speed of glider A as it passes through the light gate.   **(2)**

    ii   Glider A has a mass of 500 g and glider B has a mass of 800 g. Calculate how fast glider B will be moving, explaining your method and any assumptions you make.   **(5)**

SKILLS REASONING

e   The experiment is repeated but the two gliders are tied closer together with a shorter cotton thread. Explain what difference this would make to the result of cutting the thread, if any.   **(3)**

(Total for Question 4 = 22 marks)

HINT

Think about magnetic forces.

END OF PHYSICS ONLY

UNIT 2
ELECTRICITY

It is sometimes really difficult to imagine how we could live without electricity. As we move around we use electricity from batteries and cells for our mobile phones, mp3 players and other mobile devices. In our homes and other buildings we use electricity from the mains for heating, lighting and providing the energy for household appliances such as televisions, radios, computers and their printers. Understanding what electricity is, where it comes from and how we can control it is important if we are to make maximum use of this important source of energy.

6 MAINS ELECTRICITY

The electricity that we use for heating, lighting and air conditioning in our homes is called mains electricity and is supplied to us by power stations.

In this chapter you will learn how mains electricity is brought to our homes and supplied to our appliances. You will also read about devices that protect us from electrical shocks.

▲ Figure 6.1 Most household appliances use mains electricity as their source of energy.

LEARNING OBJECTIVES

■ Understand how the use of insulation, double insulation, earthing, fuses and circuit breakers protects the device or user in a range of domestic appliances

■ Understand why a current in a resistor results in the electrical transfer of energy and an increase in temperature, and how this can be used in a variety of domestic contexts

■ Know and use the relationship between power, current and voltage:

power = current × voltage

$P = I \times V$

and apply the relationship to the selection of appropriate fuses

■ Know the difference between mains electricity being alternating current (a.c.) and direct current (d.c.) being supplied by a cell or battery

■ Use the relationship between energy transferred, current, voltage and time:

energy transferred = current × voltage × time

$E = I \times V \times t$

UNITS

In this unit, you will need to use ampere (A) as the unit of current, coulomb (C) as the unit of charge, joule (J) as the unit of energy, ohm (Ω) as the unit of resistance, second (s) as the unit of time, volt (V) as the unit of voltage and watt (W) as the unit of power.

When you turn on your computer, television and most other appliances in your home the electricity you use is almost certainly going to come from the mains supply. This electrical energy usually enters our homes through an underground cable. The cable is connected to an electricity meter, which measures the amount of electrical energy used. From here, the cable is connected to a consumer unit or a fuse box, which contains fuses or circuit breakers for the various circuits in your home. Fuses and circuit breakers are safety devices which shut off the electricity in a circuit if the current in them becomes too large (see page 61).

Most of the wires that leave the fuse box are connected to ring main circuits that are hidden in the walls or floors around each room. Individual pieces of electrical equipment are connected to these circuits using plugs.

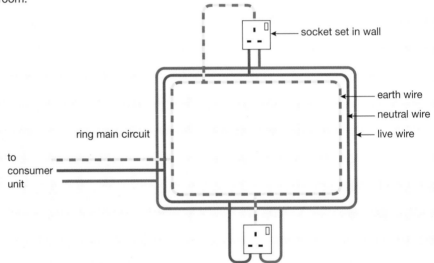

▲ Figure 6.2 Circuit breakers in a consumer unit

Ring main circuits usually consist of three wires – the live wire, the neutral wire and the earth wire.

The live wire provides the path along which the electrical energy from the power station travels. The neutral wire completes the circuit.

The earth wire usually has no current in it. It is there to protect you if an appliance develops a fault (see page 61). It provides a path for current to escape without passing through the user.

EXTENSION WORK

Ring main circuits provide a way of allowing several appliances in different parts of the same room to be connected to the mains using the minimum amount of wiring. Imagine how much wire would be needed if there was just one mains socket in each room.

socket set in wall

earth wire

neutral wire

ring main circuit

live wire

to consumer unit

▲ Figure 6.3 Ring main circuits help to cut down on the amount of wiring needed in a house.

Plugs and sockets in different countries look different, but the principles (rules) of electrical wiring are similar.

▲ Figure 6.4 Plug sockets in the UK and in Portugal

The mains electricity supplied to homes in the UK, China, India and many other countries is between 220 V and 240 V. This is a much higher **voltage** than the **cells** and batteries used in mobile electrical appliances. If you come into direct contact with mains electricity you could receive a severe electric shock, which might even be fatal. To prevent this the outer part of a plug, called the casing, is made from plastic, which is a good **insulator**.

Connections to the circuits are made via three brass pins, as the metal brass is an excellent **conductor** of electricity. Figure 6.5 shows the inside of a 3-pin plug used in the UK, but similar principles apply to all kinds of plug all over the world.

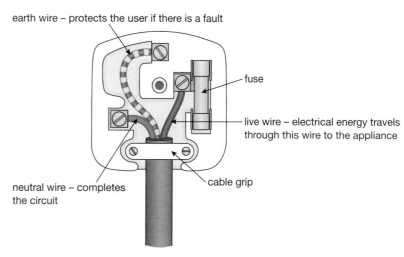

▲ Figure 6.5 A labelled plug

SAFETY DEVICES

FUSES

Many plugs contain a fuse. The fuse is usually in the form of a cylinder or cartridge, which contains a thin piece of wire made from a metal that has a low melting point. If there is too large a current in the circuit the fuse wire becomes very hot and melts. The fuse 'blows'. The circuit is now incomplete so there is no current. This prevents you getting a shock and reduces the possibility of an electrical fire. Once the fault causing the increase in current has been corrected, the blown fuse must be replaced with a new one of the same size before the appliance can be used again.

There are several sizes of fuses. The most common for domestic appliances in the UK are 3 A, 5 A and 13 A. The correct fuse for a circuit is the one that allows the correct current but blows if the current is a little larger. If the correct current in a circuit is 2 A then it should be protected with a 3 A fuse. If the correct current is 4 A then a 5 A fuse should be used. It is possible to calculate the correct size of fuse for an appliance but nowadays manufacturers provide appliances already fitted with the correct size of fuse.

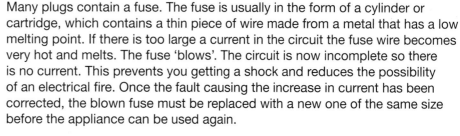

circuit symbol for a fuse

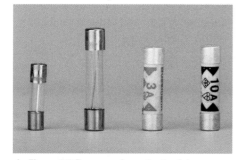

▲ Figure 6.6 Fuses are important safety devices in electrical appliances.

Modern safety devices, such as those you might find in your consumer unit, are often in the form of trip switches or circuit breakers. If too large a current flows in a circuit a switch automatically opens making the circuit incomplete. Once the fault in the circuit has been corrected, the switch is reset, usually by pressing a reset button. There is no need for the switch or circuit breaker to be replaced, as there is when fuses are used. The consumer unit shown in Figure 6.2 uses circuit breakers.

EARTH WIRES AND DOUBLE INSULATION

Many appliances have a metal casing. This should be connected to the earth wire so that if the live wire becomes damaged or breaks and comes into contact with the casing the earth wire provides a low-resistance path for the current. This current is likely to be large enough to blow the fuse and turn the circuit off. Without the earth wire anyone touching the casing of the faulty appliance would receive a severe electric shock as the current passed through them to earth (Figure 6.7).

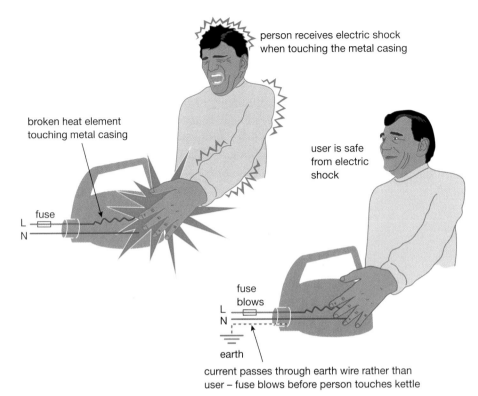

person receives electric shock when touching the metal casing

broken heat element touching metal casing

fuse

L

N

user is safe from electric shock

fuse blows

L

N

earth

current passes through earth wire rather than user – fuse blows before person touches kettle

▲ Figure 6.7 The earth wire provides protection when electrical appliances develop a fault.

▲ Figure 6.8 This plastic kettle has double insulation which means that there is no need for an earth wire.

Some modern appliances now use casings made from an insulator such as plastic rather than from metal. If all the electrical parts of an appliance are insulated in this way, so that they cannot be touched by the user, the appliance is said to have double insulation. Appliances that have double insulation use a two-wire flex. There is no need for an earth wire.

SWITCHES

Switches in mains circuits should always be placed in the live wire so that when the switch is open no electrical energy can reach an appliance. If the switch is placed in the neutral wire, electrical energy can still enter a faulty appliance, and could possibly cause an electric shock (Figure 6.9).

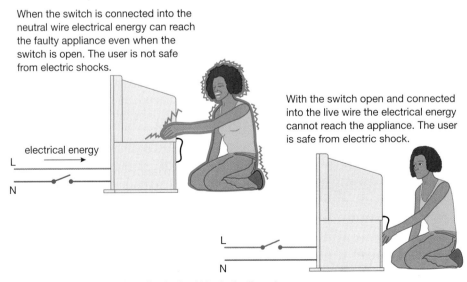

When the switch is connected into the neutral wire electrical energy can reach the faulty appliance even when the switch is open. The user is not safe from electric shocks.

electrical energy

L

N

With the switch open and connected into the live wire the electrical energy cannot reach the appliance. The user is safe from electric shock.

L

N

▲ Figure 6.9 The switch in a circuit should be in the live wire.

THE HEATING EFFECT OF CURRENT

The wiring in a house is designed to let current pass through it easily. As a result, the wires do not become warm when appliances are being used. We say that the wires have a low **resistance**. However, in some appliances, for example, kettles or toasters (Figure 6.10), we want wires (more usually called **heating elements**) to become warm. The wires of a heating element are designed to have a high resistance so that as the current passes through them energy is transferred and the element heats up. We use this heating effect of current in many different ways in our homes. You will learn more about resistance in Chapter 8.

Other common appliances that make use of the heating effect of electricity include kettles, dishwashers, electric cookers, washing machines, electric fires and hairdryers.

When current passes through the very thin wire (**filament**) of a traditional light bulb it becomes very hot and **glows** (shines) white. The bulb is transferring electrical energy to heat and light energy.

▲ Figure 6.10 The wires inside a toaster have a high resistance. They become very hot when a current passes through them.

▲ Figure 6.11 It is the heating effect of a current that is causing this bulb to glow.

ELECTRICAL POWER

Figure 6.12 shows a 50 W **halogen light bulb**. You can also buy 70 W light bulbs. Both bulbs transfer electrical energy to heat and light. The 70 W bulb will be brighter because it transfers 70 J of electrical energy every second. The dimmer 50 W halogen bulb shown transfers only 50 J of energy every second. A 70 W bulb has a higher power rating.

▲ Figure 6.12 The dimmer 50 W bulb transfers less electrical energy to heat and light energy every second.

Power is measured in joules per second or watts (W).

Devices that transfer lots of energy very quickly have their power rating expressed in kilowatts (kW).

$$1 \text{ kW} = 1000 \text{ W}$$

The power (P) of an appliance is related to the voltage (V) across it and the current (I) flowing through it.

The equation is:

power, P (watts) = current, I (amps) × voltage, V (volts)

$$P = I \times V$$

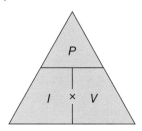

▲ Figure 6.13 You can use the triangle method for rearranging equations like $P = I \times V$.

EXAMPLE 1

A 230 V television takes a current of 3 A. Calculate the power of the television.

$P = I \times V$

$\quad = 3 \text{ A} \times 230 \text{ V}$

$\quad = 690 \text{ W}$

EXAMPLE 2

Calculate the correct fuse that should be used for a 230 V, 1 kW electric hairdryer.

$I = \dfrac{P}{V}$

$\quad = \dfrac{1000 \text{ W}}{230 \text{ V}}$

$\quad = 4.35 \text{ A}$

The correct fuse for this hairdryer is therefore a 5 A fuse.

CALCULATING THE TOTAL ENERGY TRANSFERRED BY AN APPLIANCE

The power of an appliance (P) tells you how much energy it transfers each second. This means that the total energy (E) transferred by an appliance is equal to its power multiplied by the length of time (in seconds) the appliance is being used.

energy, E (joules) = power, P (watts) × time, t (seconds)

$$E = P \times t$$

or since $P = I \times V$

$$E = I \times V \times t$$

EXAMPLE 3

Calculate the energy transferred by a 60 W bulb that is turned on for
a 20 s, and **b** 5 min.

a $E = P \times t$

     = 60 W × 20 s

     = 1200 J or 1.2 kJ

b $E = P \times t$

     = 60 W × 5 × 60 s

     = 18 000 J or 18 kJ

ALTERNATING CURRENT AND DIRECT CURRENT

If we could see the current or voltage from the mains it would appear to be
very strange. Its value increases and then decreases and then does the same
again but in the opposite direction. If we could draw these changes as a graph
they would look like a wave.

This happens because of the way in which the electricity is generated at the
power station. (see Chapter 17). A current or voltage that behaves like this is
called an alternating current (a.c.) or **alternating voltage**. This is very different
to the currents and voltages we get from batteries and cells.

Cells and batteries provide currents and voltages that are always in the same
direction and have the same value. This is called direct current (d.c.) or direct
voltage. If we drew this as a graph it would be a straight horizontal line.

Figure 6.14 shows how the voltage of an a.c. supply compares with the
voltage of a d.c. supply.

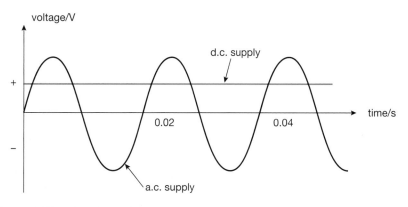

▲ Figure 6.14 How the voltage of an a.c. supply compares with that of a d.c. supply

CHAPTER QUESTIONS

More questions on domestic electricity can be found at the end of Unit 2 on page 93.

 PROBLEM SOLVING

1 a There is a current of 0.25 A in a bulb when a voltage of 12 V is applied across it. Calculate the power of the bulb.

b Calculate the voltage that is being applied across a 10 W bulb with a current of 0.2 A.

c Calculate the current in a 60 W bulb if the voltage across it is 230 V.

d How much energy is transferred if a 100 W bulb is left on for 5 hours?

 CRITICAL THINKING

2 An electric kettle is marked '230 V, 1.5 kW'.

a Explain what these numbers mean.

 PROBLEM SOLVING

b Calculate the correct fuse that should be used.

 REASONING

c Explain why a 230 V, 100 W bulb glows more brightly than a 230 V, 60 W bulb when both are connected to the mains supply.

 CRITICAL THINKING

3 a Give one advantage of using a circuit breaker rather than a wire or cartridge fuse.

b Why is the switch for an appliance always placed in the live wire?

c What is meant by the sentence 'The hairdryer has double insulation'?

SKILLS DECISION MAKING, CREATIVITY

4 Think of a room in your house where there are lots of electrical appliances. Make a list of them. Now organise your list so that the appliances that you think have the highest power rating are at the top of your list and those with the lowest are at the bottom. How could you discover if your guesses are correct?

7 CURRENT AND VOLTAGE IN CIRCUITS

We rely on electricity in many areas of our lives. This chapter looks at what electric current is. You will learn what happens to electric current in different circuits, and what effect it has.

Look around the room you are in. If you are at home, you will probably be able to see a television, a radio or a computer. If you are in a science laboratory, you may be able to see a projector, a power supply or lights in the ceiling. These and many other everyday objects need electric currents if they are to work. But what are electric currents? How are they produced and what do they do in a circuit?

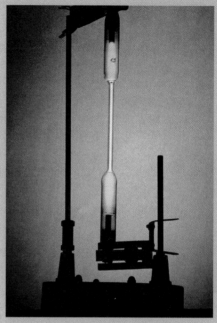

▲ Figure 7.1 The gas in this tube glows when there is a current.

LEARNING OBJECTIVES

- Explain why a series or parallel circuit is more appropriate for particular applications, including domestic lighting

- Know that electric current in solid metallic conductors is a flow of negatively charged electrons

- Know and use the relationship between voltage, current and resistance:

  voltage = current × resistance

  $V = I \times R$

- Know that current is the rate of flow of charge

- Know and use the relationship between charge, current and time:

  charge = current × time

  $Q = I \times t$

- Know that lamps and LEDs can be used to indicate the presence of a current in a circuit

- Know that the voltage across two components connected in parallel is the same

- Know that voltage is the energy transferred per unit charge passed

- Know and use the relationship between energy transferred, charge and voltage:

  energy transferred = charge × voltage

  $E = Q \times V$

- Know that the volt is a joule per coulomb

- Understand why current is conserved at a junction in a circuit

CONDUCTORS, INSULATORS AND ELECTRIC CURRENT

An electric current is a flow of charge. In metal wires the charges are carried by very small particles called electrons.

Electrons flow easily through all metals. We therefore describe metals as being good conductors of electricity. Electrons do not flow easily through plastics – they are poor conductors of electricity. A very poor conductor is known as an insulator and is often used in situations where we want to prevent the flow of charge – for example, in the casing of a plug.

In metals, some electrons are free to move between the **atoms**. Under normal circumstances this movement is random – that is, the number of electrons flowing in any one direction is roughly equal to the number flowing in the opposite direction. There is therefore no overall flow of charge.

lattice structure of metal

e⁻ e⁻ e⁻ e⁻ e⁻ e⁻

random motion of electrons

▲ Figure 7.2a With no voltage there is an equal flow of electrons in all directions.

If, however, a cell or battery is connected across the conductor, more of the electrons now flow in the direction away from the negative terminal and towards the positive terminal than in the opposite direction. We say 'there is now a net flow of charge'. This flow of charge is what we call an electric current.

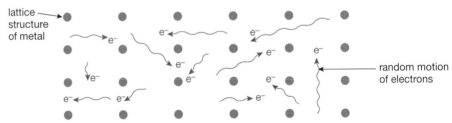

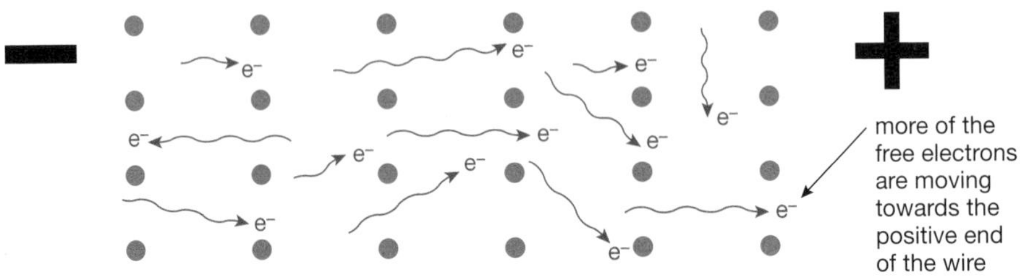

more of the free electrons are moving towards the positive end of the wire

▲ Figure 7.2b When a voltage is applied more electrons will move towards the positive.

In insulators, all the electrons are held tightly in position and are unable to move from atom to atom. Charges are therefore unable to move through insulators.

MEASURING CURRENT

EXTENSION WORK

When scientists first experimented with charges flowing through wires, they assumed that it was positive charges that were moving and that current travels from the positive to the negative. We now know that this is incorrect and that when an electric current passes through a wire it is the negative charges or electrons that move. Nevertheless when dealing with topics such as circuits and motors, it is still considered that electrons flow from positive to negative. This is conventional current.

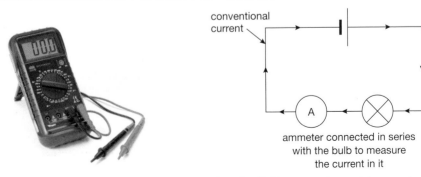

conventional current

A

ammeter connected in series with the bulb to measure the current in it

▲ Figure 7.3 An ammeter is used to measure current in a circuit. It has a very low resistance and so has almost no effect on the current.

We measure the size of the current in a circuit using an ammeter. The ammeter is connected in series (see page 72) with the part of the circuit we are interested in.

The size of an electric current indicates the rate at which the charge flows.

The charge carried by one electron is very small and would not be a very useful measure of charge in everyday life. It would be a little like asking how far away the Moon is from the Earth … and getting the answer in mm!

To avoid this problem we measure electric charge (Q) in much bigger units called coulombs (C). One coulomb of charge is equal to the charge carried by approximately six million, million, million (6×10^{18}) electrons.

We measure electric current (I) in amperes or amps (A). If there is a current of 1 A in a wire it means that 1 C of charge is passing along the wire each second.

1 C/s   =   1 A

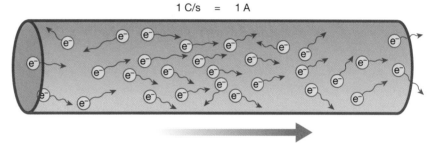

▲ Figure 7.4 One coulomb of charge flowing each second is one amp.

We can calculate the total charge that passes along a wire using the equation:

charge, Q (coulombs) = current, I (amps) × time, t (seconds)

$$Q = I \times t$$

HINT

If an examination question asks you to write out the equation for calculating charge, current or time, always give the actual equation such as $Q = I \times t$. You may not get the mark if you just draw the triangle.

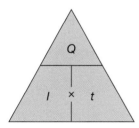

▲ Figure 7.5 You can use the triangle method for rearranging equations like this.

EXAMPLE 3

a Calculate the charge flowing through a wire in 5 s if the current is 3 A.
$Q = I \times t$
  = 3 A × 5 s
  = 15 C

b How many electrons flow through the wire in this time?
1 C of charge is carried by 6×10^{18} electrons
15 C of charge requires $15 \times 6 \times 10^{18}$ electrons = 90×10^{18} electrons
So 90 million, million, million electrons will flow along the wire in 5 s.

VOLTAGE

We often use cells or batteries to move charges around circuits. We can imagine them as being 'electron pumps'. They transfer energy to the charges. The amount of energy given to the charges by a cell or battery is measured in volts (V) and is usually indicated on the side of the battery or cell.

If we connect a 1.5 V cell into a circuit (Figure 7.6) and current flows, 1.5 J of energy is given to each coulomb of charge that passes through the cell.

If two 1.5 V cells are connected in series (Figure 7.7) so that they are pumping (pushing) in the same direction, each coulomb of charge will receive 3 J of energy.

The volt is a joule per coulomb.

▲ Figure 7.6 When one coulomb of charge passes through this cell it gains 1.5 J of energy.

▲ Figure 7.7 When one coulomb of charge passes through both these cells in turn it gains 3 J of energy.

KEY POINT

When several cells are connected together it is called a battery.

KEY POINT

Cells and batteries provide current which moves in one direction. This is known as direct current (d.c.).

As the charges flow around a circuit the energy they carry is transferred by the components they pass through. For example, when current passes through a bulb, energy is transferred to the surroundings as heat and light. When a current passes through the speaker of a radio, most of the energy is transferred as sound.

In the external part of a circuit (outside the cell or battery) the voltage across each component tells us how much energy it is transferring. If the voltage across a component is 1 V this means that the component is transferring 1 J of energy each time 1 C of charge passes through it.

We can describe the relationship between the energy transferred, charge and voltage using the equation:

energy transferred, E (joules) = charge, Q (coulombs) × voltage, V (volts)

$$E = Q \times V$$

EXAMPLE 2

The voltage across a light bulb is 12 V. Calculate the electrical energy transferred when 50 C of charge passes through it.

$E = Q \times V$

$= 50\ C \times 12\ V$

$= 600\ J$

Because $Q = I \times t$, we can also write this equation as $E = I \times t \times V$.

EXAMPLE 3

The voltage across a heater is 230 V. There is a current of 5 A through the heater for 5 mins. Calculate the total amount of energy transferred during this time.

$E = I \times t \times V$

$= 5\ A \times (5 \times 60)\ s \times 230\ V$

$= 345\,000\ J\ \text{or}\ 345\ kJ$

MEASURING VOLTAGES

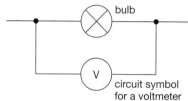

▲ Figure 7.8 A voltmeter measures voltages across a component.

We measure voltages using a voltmeter. This is connected across (in parallel with) the component we are investigating. A voltmeter connected across a cell or battery will measure the energy given to each coulomb of charge that passes through it. A voltmeter connected across a component will measure the electrical energy transferred when each coulomb of charge passes through it.

ELECTRICAL CIRCUITS

When the button on the torch shown in Figure 7.9 is pressed, the circuit is complete – that is, there are no gaps. Charges are able to flow around the circuit and the torch bulb glows. When the button is released the circuit becomes incomplete. Charges cease to flow and the bulb goes out.

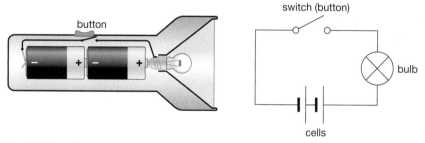

▲ Figure 7.9 A torch contains a simple electrical circuit – a series circuit.

Drawing diagrams of the actual components in a circuit is a very time-consuming and skilful task. It is much easier to use symbols for each of the components. Diagrams drawn in this way are called circuit diagrams. Figure 7.10 shows common circuit components and their symbols. You should know the common symbols but the less common ones will be given to you in the exam if you need them. Do not waste time memorising the less common ones.

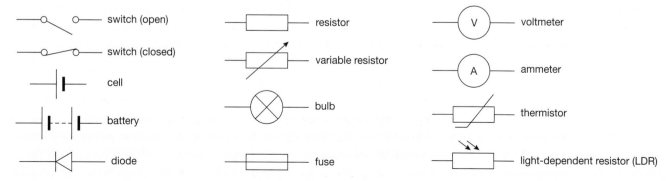

▲ Figure 7.10 Circuit symbols

▲ Figure 7.11 Glowing LEDs indicate which circuits are working

IS IT ON?

We sometimes put a small bulb or lamp in a circuit to show us if a circuit is 'turned on'. When there is a current in the circuit the bulb glows or shines. Light emitting diodes (LEDs) also glow when there is a current in a circuit but they require far less energy than bulbs. This is why many appliances such as TVs, DVD players and routers use small LEDs to show when the appliance is working or on standby.

SERIES AND PARALLEL CIRCUITS

There are two main types of electrical circuit. There are those circuits where there are no branches or junctions and there is only one path the current can follow. These simple 'single loop' circuits are called series circuits.

Circuits that have branches or junctions and more than one path that the current can follow are called parallel circuits.

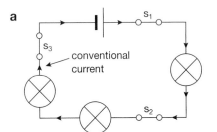
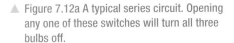

▲ Figure 7.12a A typical series circuit. Opening any one of these switches will turn all three bulbs off.

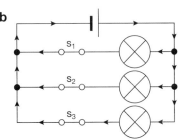

▲ Figure 7.12b A typical parallel circuit. Opening any one of these switches will turn off just the bulb in that part of the circuit.

Series circuits and parallel circuits behave differently. This makes them useful in different situations.

In a series circuit containing bulbs:

- One switch placed anywhere in the circuit can turn all the bulbs on and off.
- If any one of the bulbs breaks, it causes a gap in the circuit and all of the other bulbs will 'stop working'.
- The energy supplied by the cell is 'shared' between all the bulbs, so the more bulbs you add to a series circuit the less bright they all become.

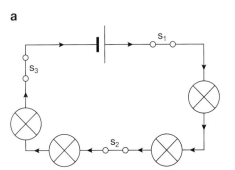

▲ Figure 7.13a Adding an extra bulb in series will result in the bulbs shining less brightly.

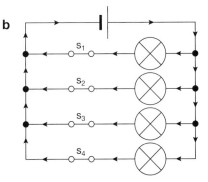

▲ Figure 7.13b Adding an extra bulb in parallel does not affect the brightness of the other bulbs.

▲ Figure 7.14 Decorative lights are usually wired in series.

In a parallel circuit containing bulbs:

■ Switches can be placed in different parts of the circuit to switch each bulb on and off individually, or all together.

■ If one bulb breaks, only the bulbs on the same branch of the circuit will be affected.

■ Each branch of the circuit receives the same voltage, so if more bulbs are added to a circuit in parallel they all keep the same brightness.

Decorative lights are often wired in series. Each bulb only needs a low voltage, so even when the voltage from the mains supply is 'shared' out between them each bulb still gets enough energy to produce light. Unfortunately, if the filament in one of the bulbs breaks then all the other bulbs will go out.

The lights in your home are wired in parallel. We know this because lights can be switched on and off separately, and the brightness of each light does not change when other lights are on or off. Also, if a bulb breaks or is removed, you can still use the other lights.

CURRENT IN A SERIES CIRCUIT

In a series circuit the current is the same in all parts; current is not used up.

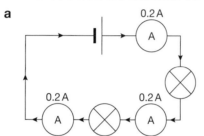

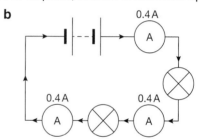

▲ Figure 7.15a In a series circuit the current does not vary.

▲ Figure 7.15b The addition of a second cell doubles the voltage applied to the circuit so the current will also double.

The size of the current in a series circuit depends on the voltage supplied to it, and the number and type of the other components in the circuit. If a second identical cell is added in series the voltage will double and so the current will also double.

CURRENT IN A PARALLEL CIRCUIT

In a parallel circuit the currents will not be the same in different parts of the circuit. The types of components in each of the different parts will affect the currents.

In a parallel circuit the number of electrons that flow into a junction each second must be equal to the number that leave each second. This means that the currents entering a junction must always be equal to those that leave. For example, in Figure 7.16 the current that enters junction P is 0.6 A. The current that leaves is 0.4 A + 0.2 A = 0.6 A.

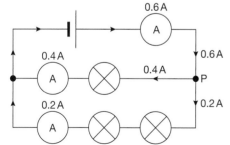

▲ Figure 7.16 There are different currents in the different parts of a parallel circuit.

CHAPTER QUESTIONS

More questions on electrical circuits can be found at the end of Unit 2 on page 93.

SKILLS CRITICAL THINKING

1 Current is a flow of charge.

a What are the charge carriers in metals?

b Explain why charges are able to flow through metals but not through a plastic.

c If the current in a heater is 3 A, calculate the charge that flows through it in:
  i 1 s

  ii 10 min
  iii 1 hour.

2 a Explain the differences between:
  i a complete circuit and an incomplete circuit

  ii a series circuit and a parallel circuit.

b Look carefully at the circuits shown below. Assuming that all switches are initially closed, decide which of the bulbs go out when each of the switches is opened in turn.

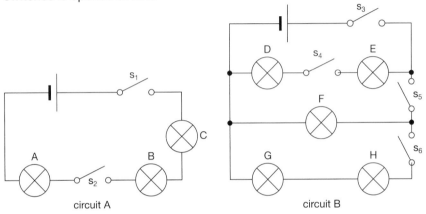

circuit A   circuit B

c In circuit A, which bulb(s) glow the brightest when all the switches in the circuit are closed?

d Explain your answer to part c.

3 The voltage between two points in a circuit is measured using a voltmeter.

a Draw a circuit diagram to show how a voltmeter should be connected to measure:
  i the voltage across a bulb   ii the voltage of a cell.

b Explain in your own words the phrase 'a cell has a voltage of 1.5 V'.

4 The diagram below shows a circuit containing two 2-way switches.

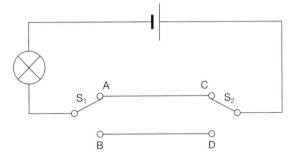

a Explain in your own words what happens when each of the switches is moved to a new position.

b Suggest one important application of this circuit in the home.

5 Should the lights for the main street in a town be wired in series or in parallel? Explain your answer.

6 Why would it not be a good idea to connect all the different parts of an electric cooker (oven, grill, heating plates) in series?

8 ELECTRICAL RESISTANCE

In this chapter you will learn what resistance is and how it can be useful in electrical appliances. You will learn what factors affect resistance and how to work out the resistance of a component by measuring the current in it and the voltage across it (Ohm's law). You will also read about some special resistors, and their uses.

It is likely that almost every day of your life you will make some adjustments to at least one electrical appliance. You may turn up the volume of your radio or change the brightness of a light. In each of these examples your adjustments are changing the currents and the voltages in the circuits of your appliance. You are doing this by altering the resistance of the circuits. This chapter will help you understand the meaning and importance of resistance and how we make use of it.

▲ Figure 8.1 Turning this dial alters the resistance in the circuit which changes the volume of the sound.

LEARNING OBJECTIVES

■ Understand how the current in a series circuit depends on the applied voltage and the number and nature of other components

■ Describe how current varies with voltage in wires, resistors, metal filament lamps and diodes, and how to investigate this experimentally

■ Describe the qualitative effect of changing resistance on the current in a circuit

■ Describe the qualitative variation of resistance of light-dependent resistors (LDRs) with illumination and thermistors with temperature

■ Calculate the currents, voltages and resistances of two resistive components connected in a series circuit

RESISTANCE

All components in a circuit offer some resistance to the flow of charge. Some (for example, connecting wires) allow charges to pass through very easily losing very little of their energy. We describe connecting wires as having very low resistance. The flow of charge through some components is not so easy and a large amount of energy may be used to move the charges through them. This energy is transferred, usually as heat. Components like these are said to have a high resistance.

We measure the resistance (R) of a component by comparing the size of the current (I) in that component and the voltage (V) applied across its ends. Voltage, current and resistance are related as follows:

voltage, V (volts) = current, I (amps) × resistance, R (ohms)

$$V = I \times R$$

We measure resistance in units called ohms (Ω).

EXAMPLE 1

When a voltage of 12 V is applied across a doorbell there is a current of 0.1 A. Calculate the resistance of the doorbell.

$$V = I \times R$$

Rearrange the equation.

$$R = \frac{V}{I}$$

$$= \frac{12\,V}{0.1\,A}$$

$$= 120\,\Omega$$

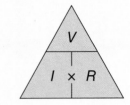

▲ Figure 8.2 You can use the triangle method for rearranging equations like $V = I \times R$.

If there are two or more resistors connected in series in a circuit, their total resistance is found by simply adding the individual resistances together. (This is not true for parallel circuits. You do not need to know how to do this.)

EXAMPLE 2

The circuit on the right contains a 12 V battery and two resistors connected in series.

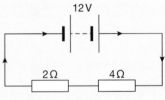

▲ Figure 8.3 Two resistor series circuit

Calculate

a the current in each of the resistors

b the voltage across each resistor.

a The total resistance the current must pass through is $2\,\Omega + 4\,\Omega = 6\,\Omega$

   The current in the circuit (I) is therefore:

$$I = \frac{V}{R}$$

$$= \frac{12\,V}{6\,\Omega}$$

$$= 2\,A$$

   The current in a series circuit is the same everywhere. So the current in both resistors is 2 A.

b Using $V = I \times R$

   for the 2 Ω resistor: $V = 2\,A \times 2\,\Omega = 4\,V$

   for the 4 Ω resistor: $V = 2\,A \times 4\,\Omega = 8\,V$

EXPERIMENT TO INVESTIGATE HOW CURRENT VARIES WITH VOLTAGE FOR DIFFERENT COMPONENTS

1 Set up the circuit shown in Figure 8.4.

2 Turn the variable resistor to its maximum value.

3 Close the switch and take the readings from the ammeter and the voltmeter.

4 Alter the value of the variable resistor again and take a new pair of readings from the meters.

5 Repeat the whole process at least six times.

6 Place the results in a table (see the table below) and draw a graph of current (*I*) against voltage (V).

> ! The resistance wire in the circuit may get hot enough to burn skin if the current/voltage is increased too much.

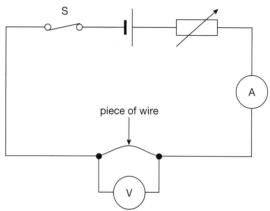

▲ Figure 8.4 This circuit can be used to investigate the relationship between current and voltage.

Current/amps	Voltage/volts
0.0	0.0
0.1	0.4
0.2	0.8
0.3	1.2
0.4	1.6
0.5	2.0

▲ Typical results table

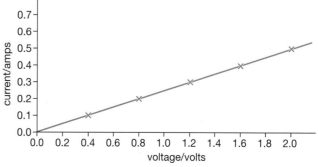

▲ Figure 8.5 Graph of results

The graph in Figure 8.5 is a straight line graph passing through the origin. The slope of the graph tells us about the resistance of the wire. The steeper the slope the smaller the resistance of the wire.

If we repeat this experiment for other components, such as a resistor, a filament bulb and a diode, the shapes of the graphs we obtain are often very different to that shown in Figure 8.5. By looking very carefully at these shapes we can see how they behave.

CURRENT/VOLTAGE GRAPH FOR A WIRE OR A RESISTOR

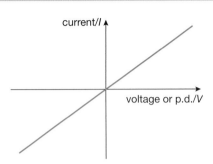

▲ Figure 8.6

The graph is a straight line. It has a constant slope. So the resistance of this component does not change.

CURRENT/VOLTAGE GRAPH FOR A FILAMENT BULB

HINT

The flatter the slope the higher the resistance.

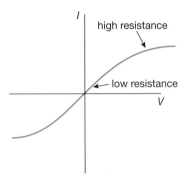

▲ Figure 8.7

This graph is not a straight line. The resistance of the bulb changes. At higher currents and voltages the slope of the graph shows us that the resistance of the filament bulb increases – that is, as the temperature of the filament increases the current decreases.

CURRENT/VOLTAGE GRAPH FOR A DIODE

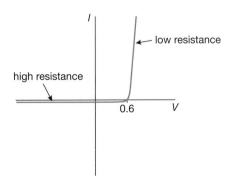

▲ Figure 8.8

This strangely shaped graph shows that diodes have a high resistance when the current is in one direction and a low resistance when it is in the opposite direction (see page 81).

USING RESISTANCE

FIXED RESISTORS

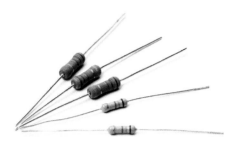

▲ Figure 8.9 A selection of resistors

In many circuits you will find components similar to those shown in Figure 8.9. They are called fixed resistors. They are included in circuits in order to control the sizes of currents and voltages. The resistor in the circuit in Figure 8.10 is included so that both the current in the bulb and the voltage applied across it are correct. Without the resistor the voltage across the bulb may cause too large a current and the bulb may 'blow' or break.

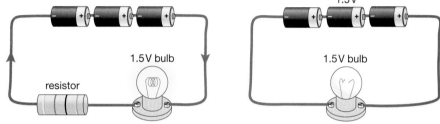

▲ Figure 8.10 The resistor in the first circuit limits the size of the current. Without the resistor the current in the second circuit is too high and the bulb breaks.

VARIABLE RESISTORS

Figure 8.11 shows examples of a different kind of resistor. They are called variable resistors as it is possible to alter their resistance. If you alter the volume of your radio using a knob you are using a variable resistor to do this.

▲ Figure 8.11 Variable resistors and their symbol

In the circuit in Figure 8.12 a variable resistor is being used to control the size of the current in a bulb. If the resistance is decreased there will be a larger current and the bulb shines more brightly. If the resistance is increased the current will be smaller and the bulb will glow less brightly or not at all. The variable resistor is behaving in this circuit as a dimmer switch. In circuits containing electric motors, variable resistors can be used to control the speed of the motor.

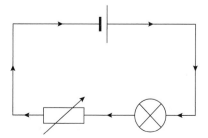

▲ Figure 8.12 Circuit with a variable resistor being used as a dimmer switch

SPECIAL RESISTORS

THERMISTORS

A **thermistor** is a resistor whose resistance changes quite a lot even with small changes in temperature.

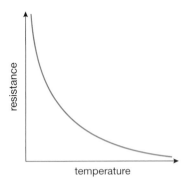

▲ Figure 8.13 A graph showing a thermistor's decreasing resistance with increasing temperature

Thermistors are used in temperature-sensitive circuits in devices such as fire alarms. They are also used in devices where it is important to make sure there is no change in temperature, for example, in freezers and computers.

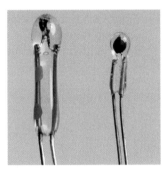

▲ Figure 8.14 Two examples of thermistors and their symbol – the resistance of a thermistor changes a lot as the temperature changes.

LIGHT-DEPENDENT RESISTORS (LDRs)

A light-dependent resistor (LDR) has a resistance that changes when light is shone on it. In the dark its resistance is high but when light is shone on it its resistance decreases.

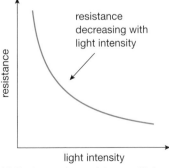

Figure 8.15 A graph showing an LDR's decreasing resistance with increasing light intensity

▲ Figure 8.16 Light-dependent resistor

LDRs are often used in light-sensitive circuits in devices such as photographic-exposure equipment, automatic lighting controls and burglar alarms.

DIODES

Diodes are very special resistors that allow charges to flow through them easily but only in one direction.

When a diode is connected as shown in circuit A in Figure 8.17, the diode offers little resistance to the charges flowing through it. But if the diode is connected the opposite way round, the diode has a very high resistance and the rate at which the charges can flow through the diode is much less – that is, the current is very small. Diodes are often used in circuits where it is important that electrons flow only in one direction. For example, they are used in **rectifier circuits** that convert alternating current into direct current. Some diodes glow when charges flow through them. They are called **light emitting diodes** (LEDs).

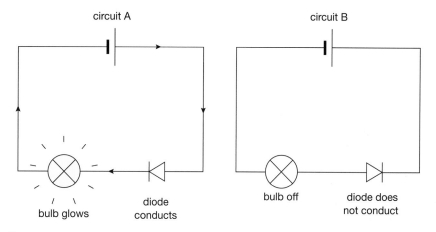

▲ Figure 8.17 Diodes will only let charges flow one way.

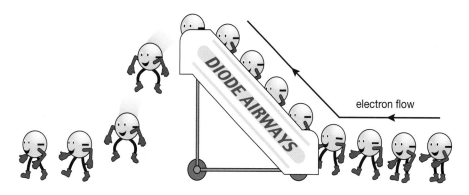

▲ Figure 8.18 Diodes are like aeroplane steps – from the ground, you can only climb them in one direction. In a diode the charge can only flow in one direction.

LOOKING AHEAD – OHM'S LAW AND TEMPERATURE

The relationship between the voltage across a component and its current is described by Ohm's law, which states:

The current in a conductor is directly proportional to the potential difference across its ends, provided its temperature remains constant.

So the resistance of a wire can be found by measuring the voltage (V) across it and the current (I) in it when this voltage is applied to the wire and then calculating a value for the **ratio** $\frac{V}{I}$ (see page 75). But the law also states that the temperature of the wire must be constant. This is because if the temperature of the wire changes, its resistance also changes.

This happens because at higher temperatures the atoms in the wire **vibrate** more vigorously, making it more difficult for the electrons to flow.

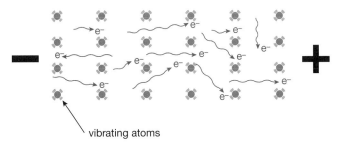

vibrating atoms

▲ Figure 8.19 At higher temperatures the increased vibration of the atoms makes it more difficult for charges to flow.

If a wire or conductor is cooled the vibration of its atoms decreases and so its resistance decreases. At very low temperatures, close to absolute zero (–273 °C), these vibrations stop and the conductor offers no resistance to the flow of charge. This event is called superconductivity and could be extremely useful. For example, when electricity flows through a superconductor there is no loss of energy. This means that by using superconductivity we could transmit electrical energy from power stations without losses. Scientists around the world are now searching for materials that are superconductors at temperatures well above absolute zero.

▲ Figure 8.20 Maglev trains use superconducting magnets to help them hover above the tracks.

More questions on electrical resistance can be found at the end of Unit 2 on page 93.

CHAPTER QUESTIONS

SKILLS   CRITICAL THINKING

SKILLS   INTERPRETATION

SKILLS   DECISION MAKING

SKILLS   INTERPRETATION

SKILLS   PROBLEM SOLVING

SKILLS   INTERPRETATION

SKILLS   CRITICAL THINKING

1 a Describe how the current in a wire changes as the voltage across the wire increases.

   b Draw a diagram of the circuit you would use to confirm your answer to part a.

   c Describe how you would use the apparatus and what readings you would take.

   d Draw an *I–V* graph for
     i a piece of wire at room temperature
     ii a filament bulb
     iii a diode.
     Explain the main features of each of these graphs.

2 a There is a current of 5 A when a voltage of 20 V is applied across a resistor. Calculate the resistance of the resistor.

   b Calculate the current when a voltage of 12 V is applied across a piece of wire of resistance 50 Ω.

   c Calculate the voltage that must be applied across a wire of resistance 10 Ω if the current is to be 3 A.

3 a Describe how the resistance of
     i a thermistor changes as its temperature changes
     ii a light-dependent resistor changes as an increasingly bright light is shone on it.
     iii Draw graphs to illustrate these changes.

   b Name one practical application for each of these resistors.

HINT

Remember when doing calculations like these to show all your working out and include units with your answer.

PHYSICS ONLY

9 ELECTRIC CHARGE

Static electricity is the result of an imbalance of charge. It can be very useful, but it can also be extremely dangerous. In this chapter, you will learn about some of the uses and problems related to static electricity.

The photograph below shows a spectacular event caused by the build up of static electricity. Flashes of lightning are seen when thunderclouds discharge their electricity. The discharging currents can be as large as 20 000 A and typically take place in just 0.1 s. Such large currents cause the air to heat up to temperatures of approximately 30 000 °C. At such high temperatures, the air immediately around the flash of lightning expands at supersonic speeds (faster than the speed of sound), causing thunder. As the sound waves travel outwards, they slow down and interact with the air and ground along their paths. This causes the continuous low rumbling sounds we hear during a thunderstorm.

▲ Figure 9.1 Lightning strikes the surface of the Earth 100 times each second. That is 8 million times a day.

LEARNING OBJECTIVES

■ Identify common materials which are electrical conductors or insulators, including metals and plastics

■ Practical: investigate how insulating materials can be charged by friction

■ Explain how positive and negative electrostatic charges are produced on materials by the loss and gain of electrons

■ Know that there are forces of attraction between unlike charges and forces of repulsion between like charges

■ Explain electrostatic phenomena in terms of the movement of electrons

■ Explain the potential dangers of electrostatic charges, e.g. when fuelling aircraft and tankers

■ Explain some uses of electrostatic charges, e.g. in photocopiers and inkjet printers

CHARGES WITHIN AN ATOM

All atoms contain small particles called protons, neutrons and electrons. The protons are found in the centre or nucleus of the atom and carry a **relative charge** of +1. The neutrons are also in the nucleus of the atom but carry no charge. The electrons travel around the nucleus in orbits. The electrons carry a relative charge of –1.

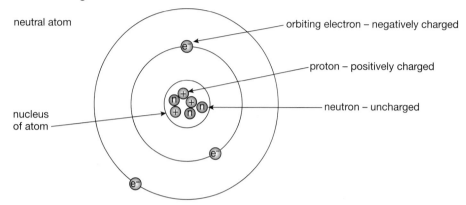

▲ Figure 9.2 A neutral atom has the same number of negative electrons and positive protons (not to scale).

Normally the number of protons in the nucleus is equal to the number of orbiting electrons. The atom therefore has no overall charge. It is neutral. If an atom gains extra electrons, it is then **negatively charged**. If an atom loses electrons, it becomes positively charged. An atom that becomes charged by gaining or losing electrons is called an ion.

CHARGING MATERIALS BY FRICTION

It is possible to charge some objects simply by rubbing them together. But these objects must be made from different materials and these materials must be **electrical insulators**.

We can test which materials are conductors and which are insulators using the circuit in Figure 9.3.

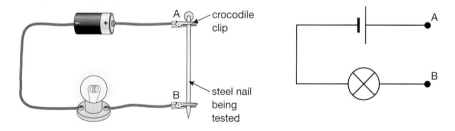

▲ Figure 9.3 Testing circuit

If we place a conductor between A and B the bulb will shine. If we place an insulator between A and B the bulb will not shine.

Materials such as graphite and metals such as copper, tin and gold allow electricity to flow through them easily. They are electrical conductors.

Materials such as plastic, rubber, glass and wood do not allow electricity to flow through them easily. They are electrical insulators.

REMINDER

It is important to remember that the rubbing action does not produce or create charge. It simply separates charge – that is, it transfers some electrons from one object to another.

ACTIVITY 1

▼ PRACTICAL: INVESTIGATE HOW INSULATORS CAN BECOME ELECTRICALLY CHARGED BY FRICTION

We can show how insulating materials can be charged by friction using an uncharged plastic rod and an uncharged cloth. First we have to check they are not already charged. We can do this by placing the rod and the cloth in turn, close to some very small pieces of paper. If the cloth or the rod is charged, some of the paper will be attracted. If they are uncharged there will be no attraction (see page 88). Now we rub the rod and the cloth together for a few seconds and then retest them. Because they are made from different insulating materials they will be charged.

If we repeat this experiment using objects that are not insulators, or are made from the same material, they will not attract the small pieces of paper. That is, they remain uncharged.

EXPLANATION

When the uncharged plastic rod is rubbed with an uncharged cloth, electrons from the atoms of the rod move onto the cloth. There is now an unequal number of positive and negative charges on each. The rod has lost electrons and so is positively charged. The cloth has gained electrons and so is negatively charged.

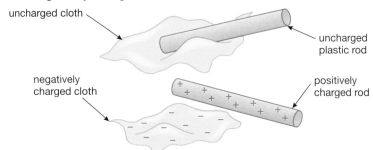

uncharged cloth

uncharged plastic rod

negatively charged cloth

positively charged rod

▲ Figure 9.4 Rubbing an uncharged rod with an uncharged piece of cloth can result in them both becoming charged.

FORCES BETWEEN CHARGES

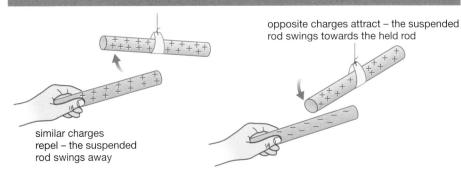

opposite charges attract – the suspended rod swings towards the held rod

similar charges repel – the suspended rod swings away

▲ Figure 9.5 Similar charges repel and opposite charges attract.

Charged objects can exert forces on other charged objects without being in contact with them. If the charges are similar, the objects repel each other. If the objects are oppositely charged they attract each other.

Static electricity is a build up of charge that does not move. The movement of charge is current electricity.

EXTENSION WORK

Although you will not be tested on the Van de Graaff generator it can be used to show dramatically the repulsion between similar charges.

Figure 9.6 shows what can happen if a person is charged with static electricity. The girl has her hands on a Van de Graaff generator. When it is turned on, charges flow onto the large metal dome. Some of the charges flow over her hands and onto all parts of her body, including her hair. Each strand (piece) of hair has the same type of charge as its neighbour. There are **repulsive forces** between the strands, so the strands repel each other. These forces cause her hair to stand on end.

For this demonstration to work, the girl must stand on an insulator to prevent any of the charges she is receiving from the generator from escaping into the floor. At the end of the demonstration the girl steps off the insulator – the charges can now escape and her hair falls. When a path is provided for charges to escape it is called earthing.

▲ Figure 9.6 A build-up of static electricity means that each hair on the girl's head has the same charge, and they repel each other.

FORCES BETWEEN CHARGED AND UNCHARGED OBJECTS

It is possible for a charged object to attract something that is uncharged. The balloon experiment demonstrates this.

If you charge a balloon by rubbing it against your jumper or your hair and then hold the balloon against a wall you will probably find that the balloon sticks to the wall. There is an attraction between the charged balloon and the uncharged wall.

EXPLANATION

After the balloon has been charged with static electricity, but before it is brought close to the wall, the charges will be distributed as shown in Figure 9.7a. The balloon is charged (we have assumed negatively charged) and the wall is uncharged – that is, it has equal numbers of positive and negative charges.

As the negatively charged balloon is brought closer to the wall some of the negative electrons are repelled from the surface of the wall. This gives the surface of the wall a slight positive charge that attracts the negatively charged balloon.

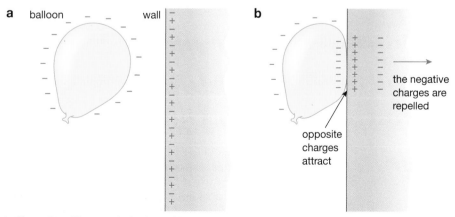

▲ Figure 9.7 **a** The negatively charged balloon approaches the neutrally charged surface of the wall.
b The positive charge on the surface of the wall attracts the oppositely charged balloon.

You can try a similar experiment using a plastic ruler and some small pieces of paper. Rub the plastic ruler on the sleeve of your jumper. The ruler will become charged. (We have assumed that the ruler has become positively charged.) If it is held close to some small, uncharged pieces of paper some electrons within the paper will be attracted to the edges closest to the ruler. There will be an attraction between these negative parts of the paper and the positive ruler.

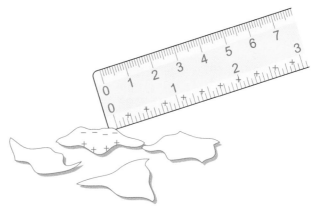

▲ Figure 9.8 The charged ruler induces a charge in the paper, and the two attract.

The charges that appear on the pieces of paper are called **induced** charges. When the ruler is removed the charges redistribute (rearrange) themselves so that the pieces of paper are once again uncharged.

EXTENSION WORK

Take a plastic ruler or rod and rub it against a jumper to charge it with static electricity. Turn on a water tap so that the water flows from the tap as slowly as possible but as a continuous flow (not as a series of drops). Hold the charged rod or ruler close to the stream of water but not in it. What happens to the water? Can you explain what is happening?

EXTENSION WORK

The gold leaf electroscope

The gold leaf electroscope is a very useful instrument for detecting charge. It is not mentioned in your examination specification but if you can understand how it works you will have understood the most important properties of static electricity.

The metal plate at the top of the electroscope is connected to a metal rod inside the case. The rod has a small thin piece of metal (called a 'leaf') attached to it. This is usually made of gold, as gold can be made into extremely thin sheets. If an electrically charged object is brought close to the plate at the top, the gold leaf will move.

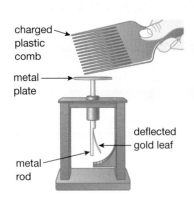

charged plastic comb

metal plate

deflected gold leaf

metal rod

▲ Figure 9.9 A modern gold leaf electroscope

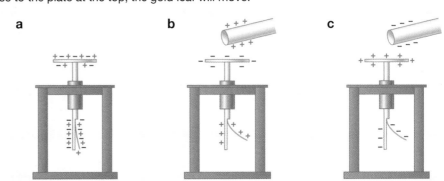

▲ Figure 9.10 **a** If there are no electrically charged objects held close to the plate the positive and negative charges are evenly spread throughout the electroscope and the gold leaf is not deflected. **b** If a positively charged object is brought close it attracts negative charges upwards onto the metal plate. This leaves both the metal rod and the gold leaf positively charged. They repel each other, so the leaf rises. **c** If a negatively charged object is brought close it repels the negatively charged electrons on the plate pushing them down onto the gold leaf and the bottom of the rod. These like charges repel and the leaf again moves upwards.

USES OF STATIC ELECTRICITY

ELECTROSTATIC PAINT SPRAYING

Painting an awkwardly shaped object, such as a bicycle frame, with a spray gun can take a long time and use a lot of paint. Using electrostatic spraying can make the process much more efficient.

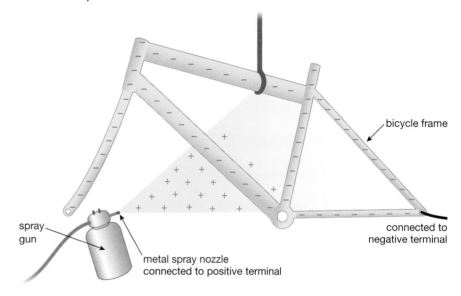

bicycle frame

spray gun

metal spray nozzle connected to positive terminal

connected to negative terminal

▲ Figure 9.11 The positive paint is attracted to all parts of the negatively charged object.

As the drops of paint emerge from the spray gun, they are charged. As the drops all carry the same charge they repel and spread out forming a thin spray. The metal bicycle frame has a wire attached to an electrical supply giving the frame the opposite charge. The paint drops are therefore attracted to the surface of the frame. There is the added benefit that paint is attracted into places, such as corners, that would normally be hard to reach.

INKJET PRINTERS

▲ Figure 9.12 An inkjet printer

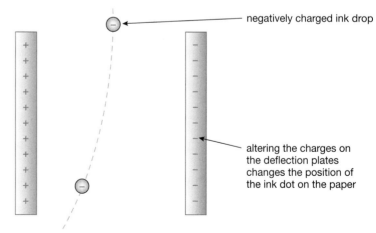

negatively charged ink drop

altering the charges on the deflection plates changes the position of the ink dot on the paper

Figure 9.13 The charged ink drops are deflected into the correct position on the paper.

Many modern **inkjet printers** use inkjets to direct a fine jet (stream) of ink drops onto paper. They do this by using electrostatic forces. Each spot of ink is given a charge so that as it falls between a pair of deflecting plates, electrostatic forces direct it to the correct position. The charges on the plates change hundreds of times each second so that each drop falls in a different position, forming pictures and words on the paper as required.

PHOTOCOPIERS

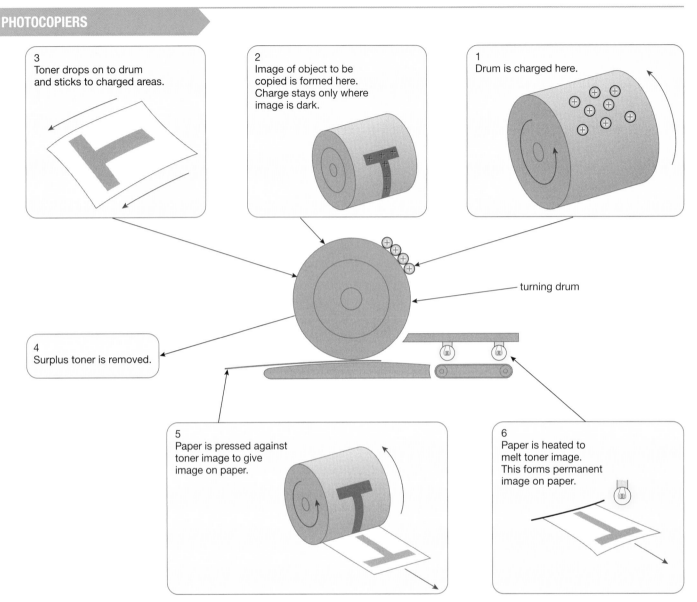

3
Toner drops on to drum and sticks to charged areas.

2
Image of object to be copied is formed here. Charge stays only where image is dark.

1
Drum is charged here.

turning drum

4
Surplus toner is removed.

5
Paper is pressed against toner image to give image on paper.

6
Paper is heated to melt toner image. This forms permanent image on paper.

▲ Figure 9.14 Static electricity is used in photocopiers.

▲ Figure 9.15 A photocopier

Positive charges are sprayed onto a turning drum whose surface is covered with a metal called selenium. A bright light is shone onto the sheet of paper to be copied. The white parts of the paper reflect light onto the drum; the dark or printed parts do not. In those places where light is reflected onto the drum the selenium loses its charge but where no light is reflected onto the drum the charge remains. A negatively charged carbon powder called toner is blown across the drum and sticks to just those parts of the drum that are charged. A sheet of paper is now pressed against the drum and picks up the pattern of the carbon powder. The powder is then fixed in place by a heater.

Many heavy industrial plants, such as steel-making furnaces and coal-fired power stations, produce large quantities of smoke. This smoke carries small particles of ash and dust into the environment, causing health problems and damage to buildings. One way of removing these pollutants from the smoke is to use electrostatic precipitators.

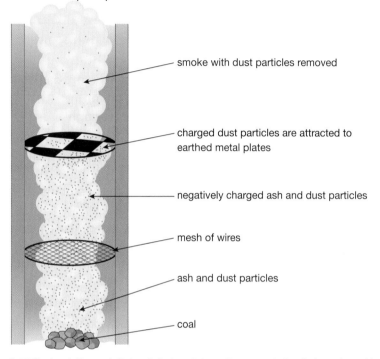

smoke with dust particles removed

charged dust particles are attracted to earthed metal plates

negatively charged ash and dust particles

mesh of wires

ash and dust particles

coal

▲ Figure 9.16 Electrostatic precipitators help to cut down the amount of pollution released into the atmosphere.

As the smoke initially rises up the chimney, it passes through a mesh of wires that are highly charged. (The wires are at a voltage of approximately −50 000 V.) As they pass through the mesh, the ash and dust particles become negatively charged. Higher up the chimney, these charged particles are attracted by and stick to, large metal earthed plates. The cleaner smoke is then released into the atmosphere. When the earthed plates are completely covered with dust and ash, they are tapped hard. The dust and ash fall into collection boxes, which are later emptied.

In a large coal-fired power station, 50–60 tonnes of dust and ash may be removed from the smoke each hour!

PROBLEMS WITH STATIC ELECTRICITY

In some situations the presence of static electricity can be a disadvantage.

As aircraft fly through the air, friction causes them to become charged with static electricity. After an aircraft has landed there is the possibility of charges escaping to earth as a spark or flash of electricity. If this takes place during refuelling, it could cause an explosion. The solution to this problem is to earth the plane with a conductor as soon as it lands and before refuelling begins, allowing the charge that has built up to flow to earth. Fuel tankers that transport fuel on roads must also be earthed before any fuel is transferred, to prevent sparks causing a fire or an explosion.

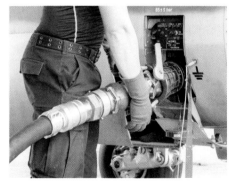

▲ Figure 9.17 Earthing during refuelling

Sometimes after a long car journey on a dry day we can become charged with static electricity and when we step from the car we might receive a small electric shock.

Our clothing can become charged with static electricity under certain circumstances. When we remove the clothes there is the possibility of receiving a small electric shock as the charges escape to earth.

CHAPTER QUESTIONS

More questions on domestic electricity can be found at the end of Unit 2 on page 93.

SKILLS CRITICAL THINKING

1 a What charge is carried by each of these particles?
   i a proton
   ii an electron
   iii a neutron

b Where inside an atom are each of the three particles mentioned in part a found?

c How many protons are there in a neutral atom compared to the number of electrons?

d What do we call an atom that has become charged by gaining or losing electrons?

SKILLS INTERPRETATION

e Describe with diagrams how two objects can be charged by friction (rubbing).

SKILLS REASONING

2 Explain the following.

a A crackling sound can sometimes be heard when removing a shirt.

b Sometimes after a journey in a car you can get a small electric shock when you touch the handle of the door.

c A plastic comb is able to attract small pieces of paper immediately after it has been used.

d After landing, aircraft are always 'earthed' before being refuelled.

SKILLS ADAPTIVE LEARNING

3 a In a photocopier, why does toner powder stick to some places on the selenium-covered drum but not to others?

SKILLS REASONING

b Explain why ash and dust particles are attracted towards the earthed metal plates of an electrostatic precipitator after they have passed through a highly negatively charged mesh of wires.

HINT

Read again about the balloon experiment.

SKILLS CRITICAL THINKING

4 Lightning is caused by clouds discharging their static electricity.

a Find out:
   i how the clouds become charged
   ii how a lightning conductor works.

b Suggest two places which might be
   i unsafe during a thunderstorm
   ii safe during a thunderstorm.

SKILLS REASONING

5 Computer chips can be damaged by static electricity if a spark jumps between a worker and a computer. Suggest how workers who build and repair computers avoid this problem.

END OF PHYSICS ONLY

HINT

See section 'Problems with static electricity'.

UNIT QUESTIONS

CRITICAL THINKING

a Which of the following is not used to protect us from the possibility of receiving an electric shock?

A double insulation

B live wire

C earth wire

D circuit breaker

(1)

b Which of the following is true for a negatively charged object?

A It will attract another negatively charged object.

B It has too few electrons.

C It has too many neutrons.

D It has gained extra electrons.

(1)

c Which of the following is true for all parallel circuits?

A Parts of the circuit can be turned off while other parts remain on.

B The current is the same in all parts of the circuit.

C There is only one path for the current to follow.

D There are no junctions or branches.

(1)

PROBLEM SOLVING

d When a voltage of 6 V is applied across a resistor there is a current of 0.1 A. The value of the resistor is

A 6 Ω

B 60 Ω

C 16.6 Ω

D 0.6 Ω

(1)

(Total for Question 1 = 4 marks)

CRITICAL THINKING  **2**

Copy and complete the following passage about electricity, filling in the spaces.

An electric current is a flow of _____. A current of 1 amp is 1 _____ of charge flowing each second. The voltage is the _____ transferred per coulomb of charge.

The current in a component depends on the voltage and the _____; the higher the resistance, the _____ the current.

(Total for Questions 2 = 5 marks)

SKILLS ANALYSIS

3 Asma set up the circuit shown below to investigate how the resistance of a bulb changes as the current in it changes.

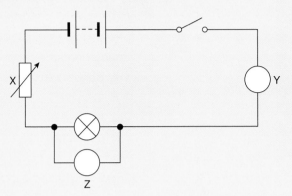

a What are the names of the instruments labelled Y and Z? **(2)**

b What is the name of the component labelled X? **(1)**

c What is the purpose of X in this circuit? **(1)**

Asma takes a series of readings. She measures the voltage across the bulb and the current in it. She then plots the graph shown below.

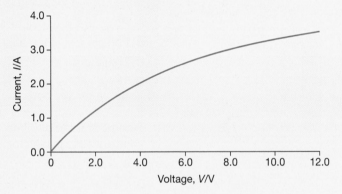

d What is the current in the bulb when a voltage of 6 V is applied across it? **(1)**

e What voltage is applied across the bulb when there is a current of 2 A? **(1)**

SKILLS PROBLEM SOLVING

f Calculate the resistance of the bulb when there is a current of 2 A. **(2)**

SKILLS REASONING

g What happens to the resistance of the bulb as the current increases? **(1)**

(Total for Question 3 = 9 marks)

SKILLS PROBLEM SOLVING

4 A simple series circuit containing a 12 V battery and a 10 Ω resistor was constructed as shown below.

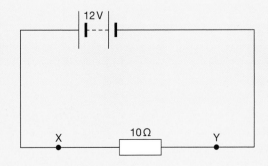

8 a Calculate the current between points X and Y. (2)

7 b Calculate the total charge that flows between X and Y in 5 s. (2)

c Calculate the energy transferred in the resistor in 1 minute. (2)

(Total for Question 4 = 6 marks)

5 An electric kettle is rated at 2 kW when connected to a 230 V electrical supply.

7 a Calculate the current when the kettle is turned on. (3)

5 b What value fuse should be included in the circuit of the kettle?
(Assume that the fuses available are 3 A, 5 A and 13 A.) (1)

SKILLS CRITICAL THINKING **6** c Modern kettles often have double insulation. Explain what this means
and how it provides extra safety for the user. (2)

SKILLS PROBLEM SOLVING **8** d Calculate the resistance of the heating element of the kettle. (3)

(Total for Question 5 = 9 marks)

SKILLS CRITICAL THINKING **6** **6** a Explain in detail how insulating materials can be charged by friction. (4)

SKILLS REASONING b When an aircraft lands it is important that it is earthed before it is
refuelled.

7 i Explain why the aircraft should be earthed. (3)

6 ii Suggest one way in which the aircraft could be earthed. (1)

7 c Explain why electrostatic painting of objects, such as bicycle frames,
makes good economic sense. (3)

SKILLS CRITICAL THINKING **6** d Describe briefly how an inkjet printer makes use of some of the
properties of static electricity. (3)

(Total for Question 6 = 14 marks)

5 **7** a Describe four uses of the heating effect of electricity in the home. (4)

6 b Explain why a double-insulated hairdryer does not need an earth wire
in its cable. (2)

(Total for Question 7 = 6 marks)

SKILLS PROBLEM SOLVING **7** **8** Calculate

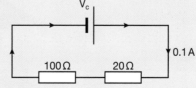

a the voltage across the 20 Ω resistor (2)

b the voltage across the 100 Ω resistor (2)

c the voltage of the cell (V_c). (1)

(Total for Question 8 = 5 marks)

UNIT 3
WAVES

There are many different types of waves. They affect all of our lives. Sometimes they are useful and can be a tremendous benefit to the way we live. Sometimes they can be dangerous and pose a real risk to life. It is therefore very important that we understand the main features and properties of waves.

10 PROPERTIES OF WAVES

Talking to someone using a mobile phone is something most of us do several times a day. The technology that had to be developed for this to happen was based on a thorough understanding of the properties of waves.

In this chapter you will learn about different types of waves and their properties (characteristics).

▲ Figure 10.1 Using microwaves to communicate

LEARNING OBJECTIVES

■ Explain the difference between longitudinal and transverse waves

■ Know the definitions of amplitude, wavefront, frequency, wavelength and period of a wave

■ Know that waves transfer energy and information without transferring matter

■ Know and use the relationship between the speed, frequency and wavelength of a wave:

wave speed = frequency × wavelength

$v = f \times \lambda$

■ Use the relationship between frequency and time period:

$$\text{frequency} = \frac{1}{\text{time period}}$$

$$f = \frac{1}{T}$$

■ Use the above relationships in different contexts including sound waves and electromagnetic waves

■ Explain that all waves can be reflected and refracted

■ Explain why there is a change in the observed frequency and wavelength of a wave when its source is moving relative to an observer, and that this is known as the Doppler effect

UNITS

In this unit, you will need to use degrees (°) as the unit of angle, hertz (Hz) as the unit of frequency, metre (m) as the unit of length, metre per second (m/s) as the unit of speed and second (s) as the unit of time.

WHAT ARE WAVES?

Waves are a way of transferring energy from place to place. As we can see in Figure 10.1 we often use them to transfer information. All these transfers take place with no matter being transferred.

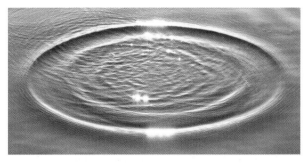

▲ Figure 10.2 Waves are produced if we drop a stone into a pond. The circular wavefronts spread out from the point of impact, carrying energy in all directions, but the water in the pond does not move from the centre to the edges.

WHAT ARE WAVEFRONTS?

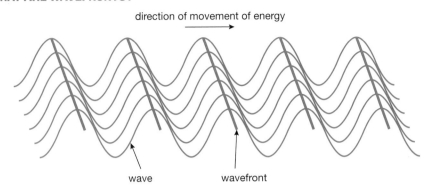

▲ Figure 10.3 Wavefronts are created by overlapping lots of different waves. A wavefront is a line where all the vibrations are in phase and the same distance from the source.

TRANSVERSE WAVES

Waves can be produced in ropes and springs. If you move one end of a spring from side to side you will see waves travelling through it. The energy carried by these waves moves along the spring from one end to the other, but if you look closely you can see that the coils of the spring are vibrating (shaking) across the direction in which the energy is moving. This is an example of a **transverse wave**.

A transverse wave is one that vibrates, or oscillates, at right angles to the direction in which the energy or wave is moving. Examples of transverse waves include light waves and waves travelling on the surface of water.

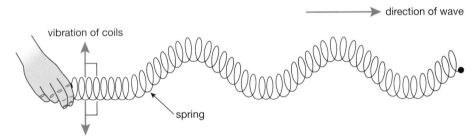

▲ Figure 10.4 A transverse wave vibrates at right angles to the direction in which the wave is moving.

LONGITUDINAL WAVES

If you push and pull the end of a spring in a direction parallel to its axis, you can again see energy travelling along it. This time however the coils of the spring are vibrating in directions that are along its length. This is an example of a **longitudinal wave**.

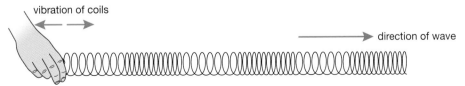

▲ Figure 10.5 A longitudinal wave vibrates along the direction in which the wave is travelling.

A longitudinal wave is one in which the vibrations, or oscillations, are along the direction in which the energy or wave is moving. Examples of longitudinal waves include sound waves.

DESCRIBING WAVES

When a wave moves through a substance, its particles will move from their equilibrium (resting position). The maximum movement of particles from their resting or equilibrium position is called its **amplitude** (A).

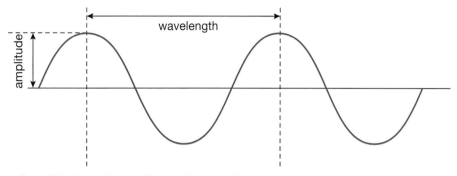

▲ Figure 10.6 A wave has amplitude and wavelength.

The distance between a particular point on a wave and the same point on the next wave (for example, from crest to crest) is called the **wavelength** (λ).

If the source that is creating a wave vibrates quickly it will produce a large number of waves each second. If it vibrates more slowly it will produce fewer waves each second. The number of waves produced each second by a source, or the number passing a particular point each second, is called the frequency of the wave (*f*). Frequency is measured in hertz (Hz). A wave source that produces five complete waves each second has a frequency of 5 Hz.

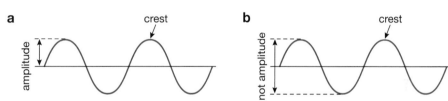

▲ Figure 10.7 The amplitude of a wave is as shown in **a**, and not as in **b**.

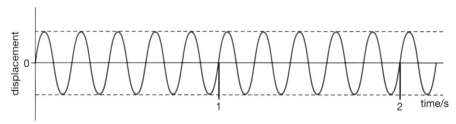

▲ Figure 10.8 This graph shows a wave with a frequency of 5 Hz.

The time it takes for a source to produce one wave is called the **time period** of the wave (*T*). It is related to the frequency (*f*) of a wave by the equation:

$$\text{frequency, } f\text{ (Hz)} = \frac{1}{\text{time period, } T\text{ (s)}}$$

$$f = \frac{1}{T}$$

This equation can also be written as

$$T = \frac{1}{f}$$

EXAMPLE 1

Calculate the period of a wave with a frequency of 200 Hz.

$$T = \frac{1}{f}$$
$$= \frac{1}{200 \text{ Hz}}$$
$$= 0.005 \text{ or } 5 \text{ ms } (1000 \text{ ms} = 1 \text{ s})$$

THE WAVE EQUATION

There is a relationship between the wavelength (λ), the frequency (f) and the wave speed (v) that is true for all waves:

wave speed, v (m/s) = frequency, f (Hz) × wavelength, λ (m)

$$v = f \times \lambda$$

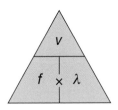

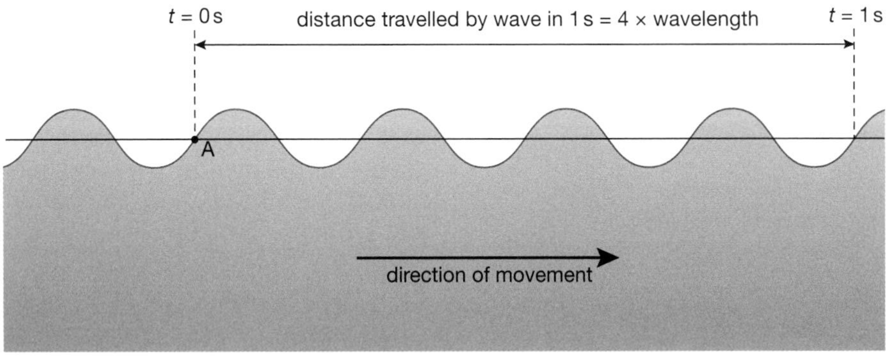

▲ Figure 10.9 You can use the triangle method for rearranging equations like $v = f \times \lambda$.

HINT

If an examination question asks you to write out the equation for calculating wave speed, wavelength or frequency, always give the actual equation such as $v = f \times \lambda$. You may not be awarded a mark if you just draw the triangle.

$t = 0$ s   distance travelled by wave in 1 s = 4 × wavelength $t = 1$ s

A

direction of movement

▲ Figure 10.10 A wave with a frequency of 4 Hz

Imagine that you have created water waves with a frequency of 4 Hz. This means that four waves will pass a particular point each second. If the wavelength of the waves is 3 m, then the waves travel 12 m each second. The speed of the waves is therefore 12 m/s.

$$v = f \times \lambda$$
$$= 4 \text{ Hz} \times 3 \text{ m}$$
$$= 12 \text{ m/s}$$

EXAMPLE 2

A **tuning fork** creates sound waves with a frequency of 170 Hz. If the speed of sound in air is 340 m/s, calculate the wavelength of the sound waves.

$$v = f \times \lambda$$

So $\lambda = \dfrac{v}{f}$

$$= \frac{340 \text{ m/s}}{170 \text{ Hz}}$$
$$= 2 \text{ m}$$

▲ Figure 10.11 A tuning fork

THE RIPPLE TANK

We can study the behaviour of water waves using a ripple tank.

When the motor is turned on, the wooden bar vibrates creating a series of ripples or wavefronts on the surface of the water. A light placed above the tank creates patterns of the water waves on the floor. By observing the patterns we can see how the water waves are behaving.

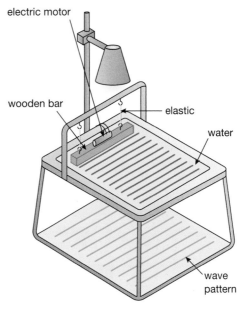

▲ Figure 10.12 The light shines through the water and we can see the patterns of the waves.

WAVELENGTH AND FREQUENCY

The motor can be adjusted to produce a small number of waves each second. The frequency of the waves is small and the pattern shows that the waves have a long wavelength.

At higher frequencies, the water waves have shorter wavelengths. The speed of the waves does not change.

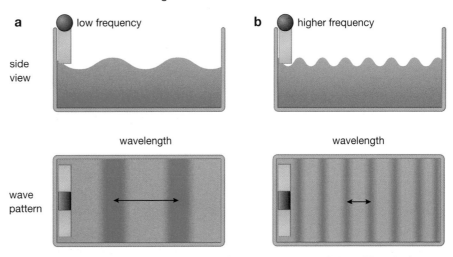

▲ Figure 10.13 When the frequency of the waves is low, the wavelength is long. When the frequency is higher, the wavelength is shorter.

REFLECTION

All waves can be reflected. If they hit a straight or flat barrier, the angle at which they leave the barrier surface is equal to the angle at which they meet the surface – that is, the waves are reflected from the barrier at the same angle as they strike it. This is described by the 'Law of Reflection' which states that:

*The **angle of incidence** is equal to the **angle of reflection**.*

KEY POINT

A normal is a line drawn at right angles to a surface.

The angle of incidence is the angle between the direction of the waves as they approach the barrier and the normal.

The angle of reflection is the angle between the direction of the waves after striking the barrier and the normal.

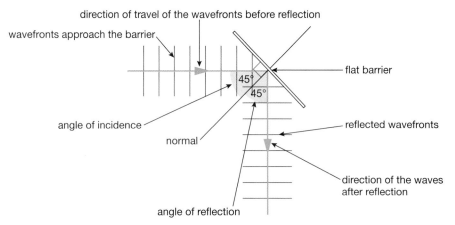

▲ Figure 10.14 Waves striking a flat barrier are reflected. The angle at which they strike the barrier is the same as the angle at which they are reflected.

EXTENSION WORK

Although you will not be asked this in your exam, it is interesting to see how waves are reflected from curved surfaces.

When the waves strike a concave barrier, they are made to converge (come together).

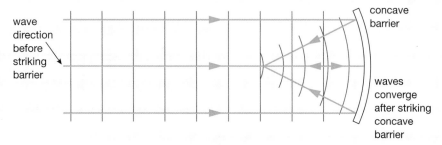

▲ Figure 10.15 Waves striking a concave barrier are reflected and converge.

▲ Figure 10.16 Radio telescopes have concave reflecting dishes so that the signals from space reflect and converge onto a detector that is placed in front of the dish.

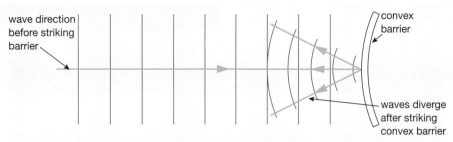

▲ Figure 10.17 Waves striking a convex barrier are reflected backwards and spread out.

When waves are reflected by a surface that is curved outwards (convex), they diverge (spread out).

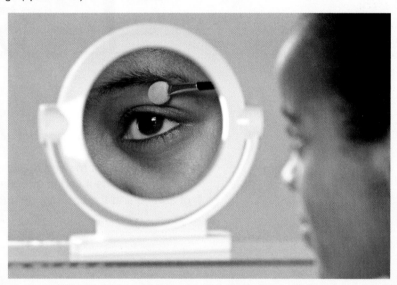

▲ Figure 10.18 The light waves reflected from this concave make-up mirror create a magnified image.

REFRACTION

The pencil in Figure 10.19 is straight but it seems to bend at the surface of the water. This happens because light waves in water travel more slowly than light waves in air. This change in speed as they leave the water causes the light waves to change direction. This change in direction is called refraction. All waves – light waves, sound waves, water waves – can be refracted.

EXTENSION WORK

Many optical instruments such as microscopes, telescopes and cameras use specially shaped pieces of glass or plastic (called lenses) to bend or refract light waves in a useful way.

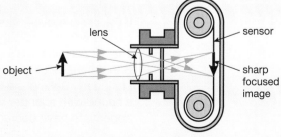

▲ Figure 10.20 In this camera, light waves are refracted by a glass lens to create a sharp image on the sensor or film. Refraction occurs because light travels more slowly in glass than in air.

a

b
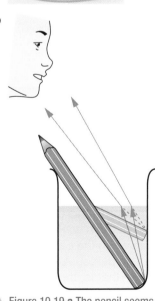

▲ Figure 10.19 **a** The pencil seems to bend at the air/water boundary. **b** Why the pencil appears to be bent. Rays of light are refracted at the water surface.

THE DOPPLER EFFECT

▲ Figure 10.21 A stationary source of sound

When a car is not moving the sound waves we receive from its engine or from its horn arrive as a series of equally (evenly) spaced wavefronts. People in front of and behind the car hear sound of the same frequency and wavelength.

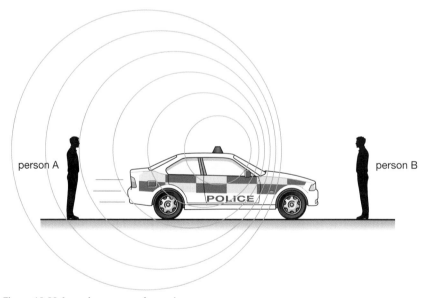

▲ Figure 10.22 A moving source of sound

If the car is moving, the wavefronts are no longer evenly spaced. Ahead of the car the wavefronts will be compressed as the car is moving in this direction. The waves will have a shorter wavelength and a higher frequency. Person B therefore hears a sound that has a higher pitch than when the car was stationary. Behind the car the waves are stretched out so person A hears a sound with a longer wavelength and lower frequency – that is, the pitch appears to have decreased.

These apparent changes in frequency, which occur when a source of waves is moving, is called the Doppler effect and is a property of all waves.

CHAPTER QUESTIONS

SKILLS   CRITICAL THINKING

SKILLS   INTERPRETATION

SKILLS   ANALYSIS, PROBLEM SOLVING

SKILLS   PROBLEM SOLVING

SKILLS   CRITICAL THINKING

SKILLS   REASONING

More questions on wave properties can be found at the end of Unit 3 on page 130.

1 a Explain the difference between a transverse wave and a longitudinal wave.

   b Give one example of each.

   c Draw a diagram of a transverse wave. On your diagram, mark the wavelength and amplitude of the wave.

2 The diagram below shows the displacement of water as a wave travels through it.

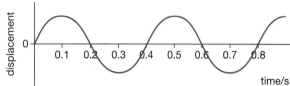

From the diagram calculate:

a the period of the wave

b the frequency of the wave.

3 The speed of sound in water is approximately 1500 m/s.

   a What is the frequency of a sound wave with a wavelength of 1.5 m?

   b What is the period of this wave?

4 a Explain why the sound produced by the horn of an approaching car seems to have a higher frequency than one that is stationary.

   b If the same car approached at a much higher speed how would this affect the frequency of the sound heard?

   c Describe the frequency of the sound heard by the observer if the car is moving away at high speed.

5 Explain why this hunter should not aim at the fish he can see.

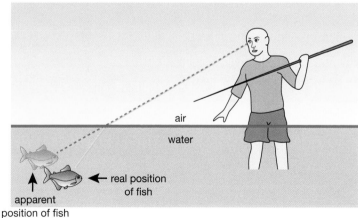

11 THE ELECTROMAGNETIC SPECTRUM

The electromagnetic spectrum is a family of waves, varying in wavelength and frequency. Although it is continuous, it is helpful to consider smaller groups of waves within the spectrum. These groups have distinct properties.
As we will see in this chapter, understanding the different properties allows us to use these waves in many situations including cooking and communication.

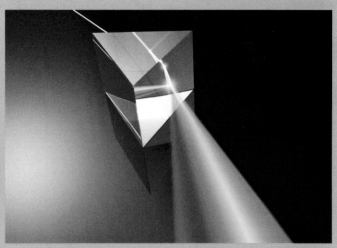

▶ Figure 11.1 When we shine white light through a prism it splits up forming a band of colours. This band is one small part of the electromagnetic spectrum.

LEARNING OBJECTIVES

- Know that light is part of a continuous electromagnetic spectrum that includes radio, microwave, infrared, visible, ultraviolet, x-ray and gamma ray radiations and that all these waves travel at the same speed in free space

- Know the order of the electromagnetic spectrum in terms of decreasing wavelength and increasing frequency, including the colours of the visible spectrum

- Explain some of the uses of electromagnetic radiations, including:

  - radio waves: broadcasting and communications

  - microwaves: cooking and satellite transmissions

  - infrared: heaters and night vision equipment

- visible light: optical fibres and photography

- ultraviolet: fluorescent lamps

- x-rays: observing the internal structure of objects and materials, including for medical applications

- gamma rays: sterilising food and medical equipment

- Explain the detrimental effects of excessive exposure of the human body to electromagnetic waves, including:

  - microwaves: internal heating of body tissue

  - infrared: skin burns

  - ultraviolet: damage to surface cells and blindness

  - gamma rays: cancer, mutation

  and describe simple protective measures against the risks

THE ELECTROMAGNETIC SPECTRUM

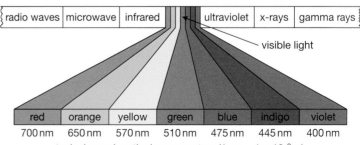

| radio waves | microwave | infrared | | ultraviolet | x-rays | gamma rays |

visible light

red	orange	yellow	green	blue	indigo	violet
700 nm	650 nm	570 nm	510 nm	475 nm	445 nm	400 nm

typical wavelengths in nanometres (1 nm = 1 × 10⁻⁹ m)

▲ Figure 11.2 The complete electromagnetic spectrum

The **electromagnetic spectrum** (EM spectrum) is a continuous spectrum of waves, which includes the visible spectrum. At one end of the spectrum the waves have a very long wavelength and low frequency, while at the other end the waves have a very short wavelength and high frequency. All the waves have the following properties:

1 They all transfer energy.
2 They are all transverse waves.
3 They all travel at 300 000 000 m/s, the speed of light in a **vacuum** (free space).
4 They can all be reflected and refracted.

Remember that the wave equations we met in the previous chapter can be applied to any member of the electromagnetic spectrum.

> **EXAMPLE 1**
>
> Yellow light has a wavelength of 5.7×10^{-7} m. What is the frequency and period of yellow light waves?
>
> $$v = f \times \lambda$$
> $$\text{So } f = \frac{v}{\lambda}$$
> $$= \frac{3 \times 10^8 \text{ m/s}}{5.7 \times 10^{-7} \text{m}}$$
> $$= 5.26 \times 10^{14} \text{ Hz}$$
>
> $$T = \frac{1}{f}$$
> $$= \frac{1}{5.26 \times 10^{14} \text{ Hz}}$$
> $$= 1.9 \times 10^{-15} \text{ s}$$

The table below shows the different groups of waves in order, and gives some of their uses.

	Typical frequency/Hz	Typical wavelength/m	Sources	Detectors	Uses
Radio waves	10^5–10^{10}	10^3–10^{-2}	radio transmitters, TV transmitters	radio and TV aerials	long-, medium- and short-wave radio, TV (UHF)
Microwaves	10^{10}–10^{11}	10^{-2}–10^{-3}	microwave transmitters and ovens	microwave receivers	mobile phone and satellite communication, cooking
Infrared (IR)	10^{11}–10^{14}	10^{-3}–10^{-6}	hot objects	skin, blackened thermometer, special photographic film	infrared cookers and heaters, TV and stereo remote controls, night vision
Visible light	10^{14}–10^{15}	10^{-6}–10^{-7}	luminous objects	the eye, photographic film, light-dependent resistors	seeing, communication (optical fibres), photography
Ultraviolet (UV)	10^{15}–10^{16}	10^{-7}–10^{-8}	UV lamps and the Sun	skin, photographic film and some fluorescent chemicals	fluorescent tubes and UV tanning lamps
X-rays	10^{16}–10^{18}	10^{-8}–10^{-10}	x-ray tubes	photographic film	x-radiography to observe the internal structure of objects, including human bodies
Gamma rays	10^{18}–10^{21}	10^{-10}–10^{-14}	radioactive materials	Geiger–Müller tube	sterilising equipment and food, radiotherapy

HINT

To remember the order of the waves in the electromagnetic spectrum try using 'Graham's Xylophone Uses Very Interesting Musical Rhythms'.

You do not need to remember the values of frequency and wavelength given in the table but you do need to know the order of the groups and which has the highest frequency or longest wavelengths. Most importantly, you need to realise that it is these differences in wavelength and frequency that give the groups their different properties – for example, gamma rays have the shortest wavelengths and highest frequencies, and carry the most energy.

RADIO WAVES

Radio waves have the longest wavelengths in the electromagnetic spectrum. They are used mainly for communication.

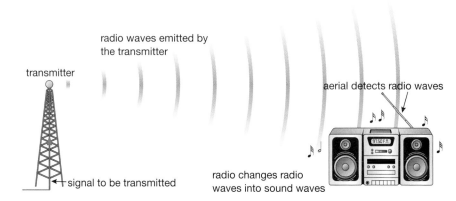

▲ Figure 11.3 Radio waves are emitted by a transmitter and detected by an aerial.

Radio waves are given out (**emitted**) by a transmitter. As they arrive at an aerial, they are detected and the information they carry can be received. Televisions and FM radios use radio waves with the shorter wavelengths to carry their signals.

MICROWAVES

Microwaves are used for communications, radar and cooking foods. Radar uses radio waves to find the position of things.

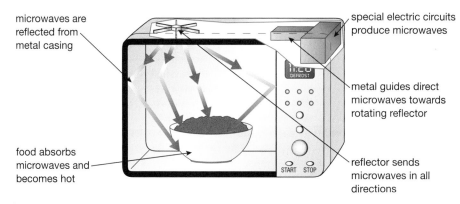

▲ Figure 11.4 Food cooks quickly in a microwave oven because water molecules in the food absorb the microwaves.

Food placed in a microwave oven cooks more quickly than in a normal oven. This is because water **molecules** in the food absorb the microwaves and become very hot. The food therefore cooks throughout rather than just from the outside.

Microwave ovens have metal screens that reflect microwaves and keep them inside the oven. This is necessary because if microwaves can cook food, they can also heat human body tissue! The microwaves used by mobile phones transmit much less energy than those used in a microwave oven, so they do not cook your brain when you use the phone. However evidence is growing that supports the idea that overuse of mobile phones may eventually harm the brain.

Microwaves are used in communications. The waves pass easily through the Earth's atmosphere and so are used to carry signals to orbiting satellites. From here, the signals are passed on to their destination or to other orbiting satellites. Messages sent to and from mobile phones are also carried by microwaves. The fact that we are able to use mobile phones almost anywhere in the home and at work confirms that microwaves can pass through glass, brick and so on.

INFRARED

All objects, including your body, emit infrared (IR) radiation. The hotter an object is, the more energy it will emit as infrared. Energy is transferred by infrared radiation to bread in a toaster or food under a grill. Electric fires also transfer heat energy by infrared.

Special cameras designed to detect infrared waves can be used to create images even when there is no visible light. These cameras have many uses, including searching for people trapped in collapsed buildings, searching for criminals and checking for heat loss from buildings.

▲ Figure 11.5 It is not possible to see these people trapped at the bottom of a cliff using normal visible light. By using infrared detectors they can be found easily and rescued.

Infrared radiation is also used in remote controls for televisions, DVD players and stereo systems. It is very convenient for this purpose because the waves are not harmful. They have a low penetrating power and will therefore operate only over small distances, so they are unlikely to interfere with other signals or waves.

The human body can be harmed by too much infrared radiation, which can cause skin burns.

▲ Figure 11.6 Signals are carried from this remote control to a TV by infrared waves.

VISIBLE LIGHT

When talking about light and colour, we often refer to the seven colours in the visible spectrum. These colours are red, orange, yellow, green, blue, indigo and violet; red light has the longest wavelength and lowest frequency. If you look back at Figure 11.1, you may only be able to make out six colours – most people have difficulty separating indigo and violet (two types of purple). Sir Isaac Newton (1642–1727) discovered that 'white' light can be split up into different colours. He believed that the number seven had magical significance, and so he decided there were seven colours in the spectrum!

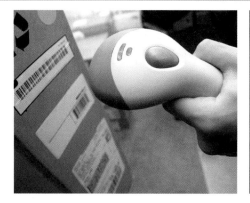

▲ Figure 11.7 Using visible light

This is the part of the electromagnetic spectrum that is visible to the human eye. We use it to see. Visible light from lasers is used to read compact discs and barcodes. It can also be sent along optical fibres, so it can be used for communication or for looking into hard-to-reach places such as inside the body of a patient (see page 121). Visible light can be detected by the sensors in digital cameras, and used to take still photographs or videos. Information stored on DVDs is also read using visible light.

ULTRAVIOLET LIGHT

▲ Figure 11.8 UV light can cause sunburn so we need to protect our skin.

Part of the light emitted by the Sun is ultraviolet (UV) light. UV radiation is harmful to human eyes and can damage the skin.

UV light causes the skin to tan, but overexposure (too much) will lead to sunburn and blistering. Ultraviolet radiation can also cause skin cancer and blindness. Protective goggles or glasses and skin creams can block the UV rays and will reduce the harmful effects of this radiation.

The **ozone layer** in the Earth's atmosphere absorbs large quantities of the Sun's UV radiation. There is real concern at present that the amount of ozone in the atmosphere is decreasing due to pollution. This may lead to increased numbers of skin cancers in the future.

Some chemicals glow (shine), or fluoresce, when under UV light. This property of UV light is used in security marker pens. The special ink is invisible in normal light but becomes visible in UV light.

▲ Figure 11.9 This red code is only visible under UV light.

mercury vapour inside the tube gives off
UV rays when a current is passed through it

when the UV light strikes the fluorescent
powder coating the tube, white light is given out

▲ Figure 11.10 Fluorescent tubes glow when UV light hits the fluorescent coating in the tube.

Fluorescent tubes glow (shine) because the UV light they produce strikes a special coating (covering) on the inside of the tube, which then emits visible light.

X-RAYS

X-rays pass easily through soft body tissue but cannot pass through bones. As a result, radiographs or x-ray pictures can be taken to check a patient's bones.

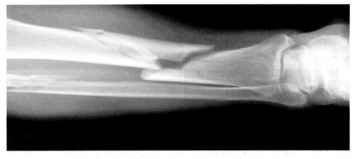

▲ Figure 11.11 X-ray of a broken leg

Working with x-rays can cause cancer. Radiographers, who take x-rays, are at risk and have to stand behind lead screens or wear protective clothing.

X-rays are also used in industry to check the internal structures of objects – for example, to look for cracks and faults in buildings or machinery – and at airports as part of the security checking procedure.

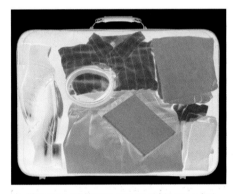

▲ Figure 11.12 X-rays were used to see what was in this suitcase.

GAMMA RAYS

Gamma rays, like x-rays, are highly penetrating rays and can cause damage to living cells. The damage can cause mutations (negative changes), which can lead to cancer. They are used to **sterilise** medical instruments, to kill microorganisms so that food will keep for longer and to treat cancer using radiotherapy. Gamma rays can both cause and cure cancer. Large doses of gamma rays targeted directly at the cancerous growth can be used to kill the cancer cells completely.

Like x-rays the use of lead screens, boxes and aprons can prevent the damage caused by gamma rays (overexposure).

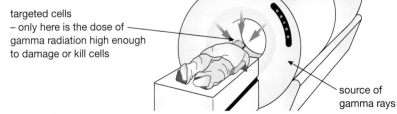

targeted cells
– only here is the dose of
gamma radiation high enough
to damage or kill cells

source of
gamma rays

▲ Figure 11.13 The gamma rays are aimed carefully so that they cross at the exact location of the cancerous cells.

CHAPTER QUESTIONS

More questions on using waves can be found at the end of Unit 3 on page 130.

SKILLS   CRITICAL THINKING

1 a Name four wave properties that are common to all members of the electromagnetic spectrum.

  b Name three types of wave that can be used for communicating.

  c Name two types of wave that can be used for cooking.

  d Name one type of wave that is used to treat cancer.

  e Name one type of wave that might be used to 'see' people in the dark.

  d Name one type of wave that is used for radar.

SKILLS   REASONING

2 Explain why:

  a microwave ovens cook food much more quickly than normal ovens

  b x-rays are used to check for broken bones

  c it is important not to damage the ozone layer around the Earth

  d food stays fresher for longer after it has been exposed to gamma radiation.

3 a Explain one way in which you could prevent overexposure (damage) by the following waves:
    i x-rays
    ii ultraviolet waves.

SKILLS   CRITICAL THINKING

  b Select one of the above waves and then describe one consequence of overexposure.

SKILLS   INTERPRETATION

4 Copy and complete the table below for four more different wave groups within the electromagnetic spectrum.

Type of radiation	Possible harm	Precautions
x-rays	cancer	lead screening

12 LIGHT WAVES

We see objects because they emit or reflect light. In this chapter you will learn how light behaves when it reflects from different surfaces, and what happens when light travels from one transparent material to another.

▲ Figure 12.1 In the Hall of Mirrors at the fairground the reflection of light can be very confusing!

LEARNING OBJECTIVES

■ Know that light waves are transverse waves and that they can be reflected and refracted

■ Use the law of reflection (the angle of incidence equals the angle of reflection)

■ Draw ray diagrams to illustrate reflection and refraction

■ Practical: investigate the refraction of light, using rectangular blocks, semi-circular blocks and triangular prisms

■ Know and use the relationship between refractive index, angle of incidence and angle of refraction:

$$n = \frac{\sin i}{\sin r}$$

■ Practical: investigate the refractive index of glass, using a glass block

■ Describe the role of total internal reflection in transmitting information along optical fibres and in prisms

■ Explain the meaning of critical angle c

■ Know and use the relationship between critical angle and refractive index:

$$\sin c = \frac{1}{n}$$

SEEING THE LIGHT

The patient shown in Figure 12.2 has a cataract. The front of one of his eyes has become so cloudy that he is unable to see. Nowadays it is possible to remove this damaged part of the eye and replace it with a clear plastic that will allow light to enter the eye again.

▲ Figure 12.2 Cataracts mean that light cannot enter the eye correctly.

There are many sources of light, including the Sun, the stars, fires, light bulbs and so on. Objects such as these that emit their own light are called luminous objects. When the emitted light enters our eyes we see the object. Most objects, however, are non-luminous. They do not emit light. We see these non-luminous objects because of the light they reflect.

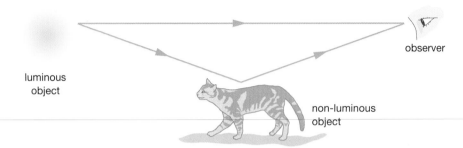

▲ Figure 12.3 Luminous objects, such as the Sun, give out light. Non-luminous objects only reflect light.

REFLECTION

When a ray of light strikes a plane (flat) mirror, it is reflected so that the angle of incidence (i) is equal to the angle of reflection (r).

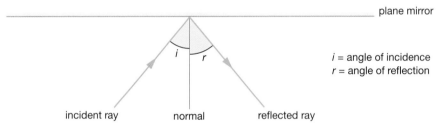

▲ Figure 12.4 Light is reflected from a plane mirror. The angle of incidence is equal to the angle of reflection. The normal is a line at right angles to the mirror.

Mirrors are often used to change the direction of a ray of light. One example of this is the simple periscope, which uses two mirrors to change the direction of rays of light.

Rays from the object strike the first mirror at an angle of 45° to the normal. The rays are reflected at 45° to the normal and so are turned through an angle of 90° by the mirror. At the second mirror the rays are again turned through 90°. Changing the direction of rays of light in this way allows an observer to use a periscope to see over or around objects.

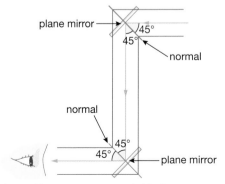

▲ Figure 12.5 A periscope is used to see over or around objects.

REFRACTION

▲ Figure 12.6 This rainbow is caused by refraction.

Rays of light can travel through many different transparent media, including air, water and glass. Light can also travel through a vacuum. In a vacuum and in air, light travels at a speed of 300 000 000 m/s. In other media it travels more slowly. For example, the speed of light in glass is approximately 200 000 000 m/s. When a ray of light travels from air into glass or water it slows down as it crosses the border between the two media. This change in speed may cause the ray to change direction. This change in direction of a ray is called refraction.

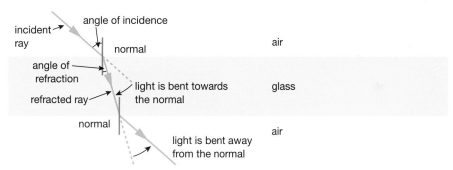

▲ Figure 12.7 This light ray is being refracted twice – once as it travels from air into glass and then as it travels from glass to air.

As a ray enters a glass block, it slows down and is refracted towards the normal. As the ray leaves the block it speeds up and is refracted away from the normal.

If the ray strikes the boundary between the two media at 90°, the ray continues without change of direction (Figure 12.8).

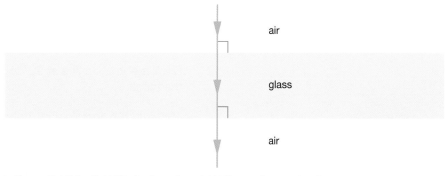

▲ Figure 12.8 If the light hits the boundary at 90° the ray does not bend.

REFRACTIVE INDEX

Different materials can bend rays of light by different amounts. We describe this by using a number called the **refractive index** (n). The refractive index of glass is about 1.5 and water is 1.3. This tells us that under similar circumstances glass will refract light more than water.

We can use the equation below to calculate the refractive index of a material:

$$n = \frac{\sin i}{\sin r}$$

where i is the angle of incidence and r is the **angle of refraction**.

EXAMPLE 1

In an experiment similar to the one shown in Figure 12.9, the angle of incidence was measured as 30° and the angle of refraction as 19°. Calculate the refractive index of the glass block.

$$n = \frac{\sin i}{\sin r}$$

$$= \frac{\sin 30}{\sin 19}$$

$$= \frac{0.5}{0.33}$$

$$= 1.52$$

! Ray box lamps get hot enough to burn skin and char paper. Glass blocks and prisms should be handled carefully and not knocked together – they can splinter or shatter.

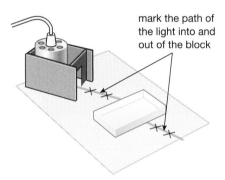

mark the path of the light into and out of the block

ACTIVITY 1

▼ PRACTICAL: INVESTIGATE THE REFRACTIVE INDEX FOR GLASS

You can investigate the refractive index of glass using a ray box and a rectangular glass block.

1 Shine a ray of light onto one of the sides of the glass block, so that the ray emerges on the opposite side of the block. Mark the directions of both of these rays with crosses.

2 Draw around the glass block before removing it.

3 Using the crosses, draw in the direction of both rays.

4 Draw in the direction of the ray that travelled inside the glass block.

5 Draw a normal (a line at 90° to the glass surface) where the ray enters the block.

6 Measure the angles of incidence (i) and refraction (r) (see Figure 12.9).

7 Use the equation $n = \frac{\sin i}{\sin r}$ to find the refractive index of the glass block.

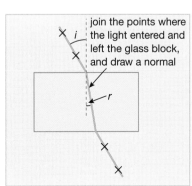

join the points where the light entered and left the glass block, and draw a normal

▲ Figure 12.9 How to investigate refraction using a rectangular glass block

TOTAL INTERNAL REFLECTION

When a ray of light with a small angle of incidence passes from glass into air, most of the light is refracted away from the normal but if we look carefully we can see that there is a small amount that is reflected from the boundary. Total internal reflection only occurs when rays of light are travelling towards a boundary with a less optically dense medium (a medium with a lower refractive index).

The symbol 'r' is used for both the angle of reflection and the angle of refraction.

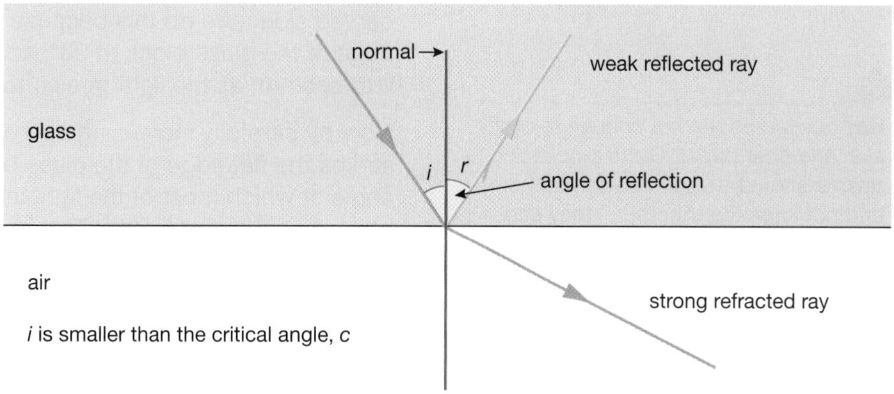

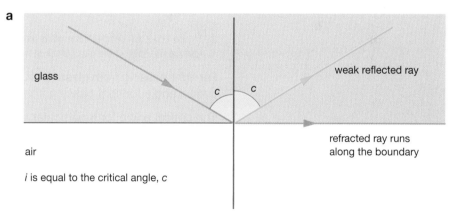

▲ Figure 12.10 A ray of light travelling from glass to air

But as the angle of incidence in the glass increases, the angle of refraction also increases until it reaches a special angle called the critical angle (c). The angle of refraction now is 90°.

The critical angle is the smallest possible angle of incidence at which light rays are totally internally reflected.

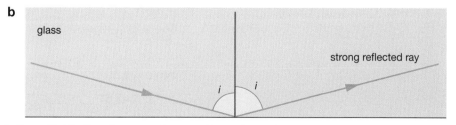

▲ Figure 12.11a Ray of light strikes glass/air boundary at the critical angle

▲ Figure 12.11b When i is greater than c total internal reflection occurs.

When i is greater than the critical angle, all the light is reflected at the boundary. No light is refracted. The light is totally internally reflected.

ACTIVITY 2

▼ PRACTICAL: INVESTIGATE TOTAL INTERNAL REFLECTION

You can investigate total internal reflection in the laboratory using a semi-circular glass block and a ray box. As shown in Figure 12.12a, a ray of light is directed at the centre of the straight side of the block through the curved side. (We do this because the incident ray will then always hit the edge of the glass block at 90°, so there are no refraction effects to take into account as the light goes into the block.)

Now by carefully increasing and decreasing the angle at which the ray strikes the flat edge of the glass block, we can discover the smallest angle at which most of the light is refracted along the edge of the glass block (see Figure 12.12b). This angle is the critical angle.

Ray box lamps get hot enough to burn skin and char paper. Glass blocks and prisms should be handled carefully and not knocked together – they can splinter or shatter.

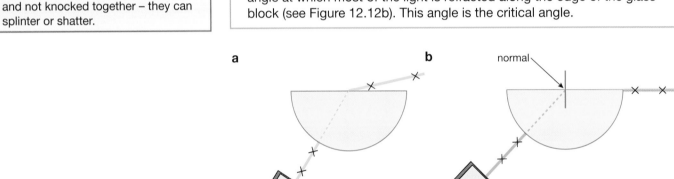

▲ Figure 12.12 **a** A semi-circular glass block used to demonstrate total internal reflection **b** Light striking the edge of the glass block at the critical angle

For light passing from glass to air, the critical angle is typically 42° and the critical angle for light passing from water to air is 49°.

The critical angle for a particular medium is related to its refractive index by this equation:

$$\sin c = \frac{1}{n}$$

EXAMPLE 2

The refractive index for a type of glass is 1.45. Calculate the critical angle.

$$\sin c = \frac{1}{n}$$

$$\sin c = \frac{1}{1.45}$$

$$\sin c = 0.69$$

$$c = 43.6°$$

We sometimes use **prisms** rather than mirrors to reflect light. The light is totally internally reflected by the prism.

Ray box lamps get hot enough to burn skin and char paper. Glass blocks and prisms should be handled carefully and not knocked together – they can splinter or shatter.

ACTIVITY 3

▼ PRACTICAL: INVESTIGATE TOTAL INTERNAL REFLECTION IN PRISMS

If you shine a ray of light into a prism as shown in Figure 12.13 it will strike the far surface at an angle of 45°. The critical angle for glass is about 42° so the ray will be totally internally reflected. You will see therefore that the ray will be reflected through an angle of 90°.

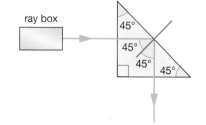

▲ Figure 12.13 Turning through 90° using total internal reflection

If you shine a ray into the prism as shown in Figure 12.14 the ray will be reflected through an angle of 180° – that is, it will go back in the direction from which it came.

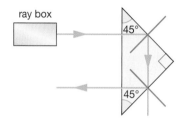

▲ Figure 12.14 Turning through 180° using total internal reflection

USING TOTAL INTERNAL REFLECTION

THE PRISMATIC PERISCOPE

The images produced by prisms are often brighter and clearer than those produced by mirrors. A periscope that uses prisms to reflect the light is called a prismatic periscope. Light passes through the surface AB of the first prism at 90° and so does not change direction (it is undeviated). It then strikes the surface AC of the prism at an angle of 45°. The critical angle for glass is 42° so the ray is totally internally reflected and is turned through 90°. When it leaves the first prism the light travels to a second prism. The second prism is positioned so that the ray is again totally internally reflected. The ray emerges parallel to the direction in which it was originally travelling.

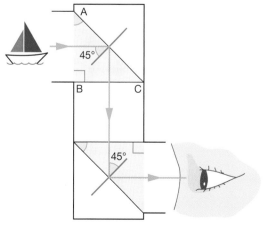

▲ Figure 12.15 Total internal reflection in a prismatic periscope

BICYCLE AND CAR REFLECTORS

▲ Figure 12.17 Reflectors like these can save lives.

a

b

after total internal reflection, the light travels back towards the source (for example, car headlights)

bicycle reflector

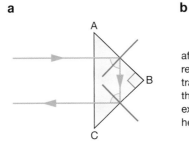

▲ Figure 12.16 Prisms can also be used as reflectors.

Light entering the prism in Figure 12.16 is totally internally reflected twice. It emerges from the prism travelling back in the direction from which it originally came. This arrangement is used in bicycle or car reflectors.

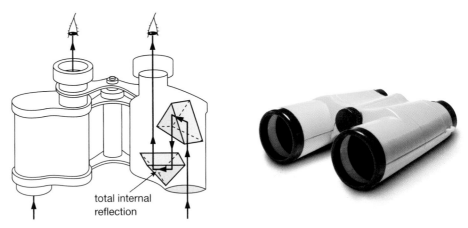

total internal reflection

▲ Figure 12.18 Total internal reflection inside binoculars   ▲ Figure 12.19 Prismatic binoculars

Binoculars also make use of total internal reflection within prisms.

Each side of a pair of binoculars contains two prisms to totally internally reflect the incoming light. Without the prisms, binoculars would have to be very long to obtain large **magnifications** and would look like a pair of telescopes.

OPTICAL FIBRES

One of the most important **applications** for total internal reflection is the optical fibre. This is a very thin piece of fibre composed of two different types of glass. The centre is made of a glass that has a high refractive index surrounded by a different type of glass that has a lower refractive index.

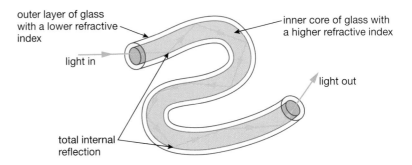

outer layer of glass with a lower refractive index

inner core of glass with a higher refractive index

light in

light out

total internal reflection

▲ Figure 12.20 In an optical fibre, light undergoes total internal reflection.

▲ Figure 12.21 Optical fibres

As the fibres are very narrow, light entering the inner core always strikes the boundary of the two glasses at an angle that is greater than the critical angle. No light escapes across this boundary. The fibre therefore acts as a 'light pipe' providing a path that the light follows even when the fibre is curved.

Large numbers of these fibres fixed together form a bundle. Bundles can carry sufficient light for images of objects to be seen through them. If the fibres are tapered (narrower at one end) it is also possible to produce a magnified image.

Figure 12.22 shows optical fibres in an endoscope. The endoscope is used by doctors to see the inside the body – for example, to examine the inside of the stomach. Endoscopes can also be used by engineers to see hard-to-reach parts of machinery.

Light travels down one bundle of fibres and shines on the object to be viewed. Light reflected by the object travels up a second bundle of fibres. An image of the object is created by the eyepiece.

THE ENDOSCOPE

By using optical fibres to see what they are doing, doctors can carry out operations through small holes made in the body, rather than through large cuts. This is called 'keyhole surgery'. This is less stressful for patients and usually leads to a more rapid recovery.

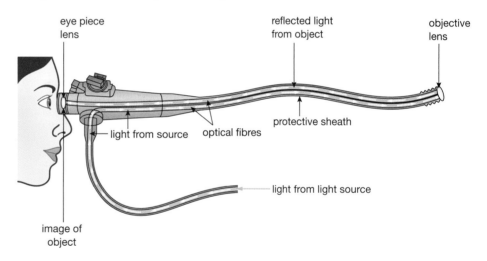

▲ Figure 12.22 Optical fibres are used in endoscopes to see inside the body.

OPTICAL FIBRES IN TELECOMMUNICATIONS

Modern telecommunications systems use optical fibres rather than copper wires to transmit messages as less energy is lost. Electrical signals from a telephone are converted into light energy produced by tiny lasers, which send pulses (small amounts) of light into the ends of optical fibres. A light-sensitive detector at the other end changes the pulses back into electrical signals, which then flow into a telephone receiver (ear piece).

CHAPTER QUESTIONS

SKILLS INTERPRETATION, PROBLEM SOLVING

SKILLS INTERPRETATION

SKILLS CRITICAL THINKING

SKILLS INTERPRETATION

SKILLS PROBLEM SOLVING

SKILLS INTERPRETATION

SKILLS CRITICAL THINKING

SKILLS INTERPRETATION

SKILLS CRITICAL THINKING

SKILLS INTERPRETATION

SKILLS REASONING

SKILLS INTERPRETATION

SKILLS INTERPRETATION, CRITICAL THINKING

SKILLS CRITICAL THINKING

More questions on refraction can be found at the end of Unit 3 on page 130.

1 Draw a ray diagram to show how a ray of light can be turned through 100° using two plane mirrors. Mark on your diagram a value for the angle of incidence at each of the mirrors.

2 a Draw a diagram to show the path of a ray of light travelling from air into a rectangular glass block at an angle of about 45°.

 b Show the path of the ray as it emerges from the block.

 c Explain why the ray changes direction each time it crosses the air/glass boundary.

 d Draw a second diagram showing a ray that travels through the block without its direction changing.

3 In an experiment to measure the refractive index of a type of glass, the angle of refraction was found to be 31° when the angle of incidence was 55°.

 a Calculate the refractive index of the glass.

 b What would the angle of refraction be for a ray with an angle of incidence of 45°?

 c Calculate the critical angle for the glass.

4 a Draw a diagram to show how a prism can create a rainbow of colours.

 b Explain how these colours are produced by the prism.

5 Draw three ray diagrams to show what happens to a ray of light travelling in a glass block in the following situations. It hits a face of the block at an angle:

 a less than the critical angle

 b equal to the critical angle

 c greater than the critical angle.

6 a What is meant by 'total internal reflection of light' and under what conditions does it occur?

 b Draw a diagram to show how total internal reflection takes place in a prismatic periscope.

 c Give one advantage of using prisms in a periscope rather than plane mirrors.

 d Draw a second diagram to show how a prism could be used to turn a ray of light through 180°. Give one application of a prism used in this way.

7 a Explain why a ray of light entering an optical fibre is unable to escape through the sides of the fibre. Include a ray diagram in your explanation.

 b Explain how doctors use optical fibres to see inside the body.

 c Name one other use of optical fibres.

13 SOUND

Sound waves are longitudinal waves and not transverse waves like light. Nevertheless they can be reflected and refracted in just the same way. In this chapter you will learn about the nature and behaviour of sound waves, and how we make use of them in our everyday lives ... and not just when we talk!

▲ Figure 13.1 Using sound waves to communicate.

LEARNING OBJECTIVES

- Know that sound waves are longitudinal waves which can be reflected and refracted

PHYSICS ONLY

- Know that the frequency range for human hearing is 20–20 000 Hz
- Practical: investigate the speed of sound in air

- Understand how an oscilloscope and microphone can be used to display a sound wave
- Practical: investigate the frequency of a sound wave using an oscilloscope
- Understand how the pitch of a sound relates to the frequency of vibration of the source
- Understand how the loudness of a sound relates to the amplitude of vibration of the source

Figure 13.2 shows part of the sound system used by a band playing at a concert. This equipment must produce sounds that are loud enough to be heard by all the audience and the sound quality must be good enough for the music to be appreciated. In this chapter we are going to look at how sounds are made and how they travel as waves.

SOUND WAVES

Sounds are produced by objects that are vibrating. We hear sounds when these vibrations, travelling as sound waves, reach our ears.

▲ Figure 13.2 The sound produced by the speakers must be loud but also of good quality.

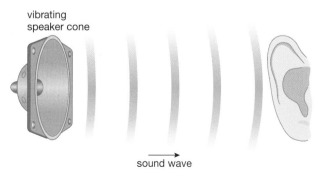
vibrating speaker cone

sound wave

▲ Figure 13.3 The loudspeaker vibrates and produces sound waves.

You will not be tested on this more detailed description of how sounds from a loudspeaker are heard. As the speaker cone moves to the right, it pushes air molecules closer together, creating a compression. These particles then push against neighbouring particles so that the compression appears to be moving to the right. Behind the compression as the speaker cone moves to the left is a region where the particles are spread out. This region is called a rarefaction. After the cone has vibrated several times, it has created a series of compressions and rarefactions travelling away from it. This is a longitudinal sound wave (see page 98). When the waves enter the ear, they strike the eardrum and make it vibrate. These vibrations are changed into electrical signals, which are then detected by the brain.

PHYSICS ONLY

MEASURING THE SPEED OF SOUND

ACTIVITY 1

▼ PRACTICAL: INVESTIGATE THE SPEED OF SOUND USING ECHOES

You can measure the speed of sound using two bits of wood and a stopwatch. Stand some distance away from a friend, start the stopwatch when you see her bang the two bits of wood together and stop the stopwatch when you hear the sound. Now using the equation

$$\text{speed}, v = \frac{\text{distance}, s}{\text{time}, t}$$ you can calculate the speed of sound. The answer

you are likely to get will not be very accurate unless you have very good eyesight and very fast reactions.

You can obtain a more accurate measurement using echoes (sounds that are reflected off a surface).

1　Both you and your friend should stand at least 50 m away from a large wall or building.

2　Bang the two pieces of wood together and listen for the echo.

3　Bang the pieces of wood together each time you hear the echo. This will create a regular rhythm of claps.

4　Ask the friend to time you doing 20 claps. During this time the sound will have travelled, for example, 50 m × 20 × 2 (to the wall and back 20 times), and you can divide this distance by the time to work out the speed of the sound.

! A hinged 'clapper board' works best but it needs exterior handles to prevent fingers being trapped.

The speed of sound in air is approximately 340 m/s, although this value does vary a little with temperature.

EXAMPLE 1

A girl stands 100 m from a tall building clapping her hands each time she hears an echo of her clap. It takes 11.7 s for her to hear 20 echoes. Calculate the speed of sound.

$$v = \frac{s}{t}$$

$$= \frac{100\,\text{m} \times 2 \times 20}{11.7\,\text{s}}$$

$$= 342 \text{ m/s}$$

END OF PHYSICS ONLY

REFLECTION

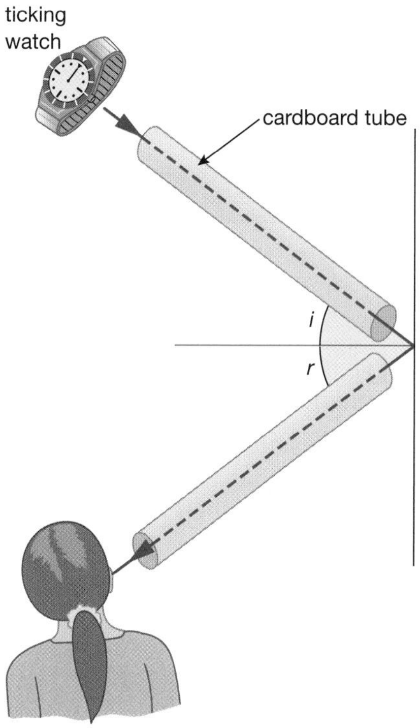

Sound waves behave in the same way as any other wave.

When a sound wave strikes a surface it may be reflected. Like light waves, sound waves are reflected from a surface so that the angle of incidence is equal to the angle of reflection (see Figure 13.4).

Ships often use echoes to discover the depth of the water beneath them. This is called echo sounding.

1 Sound waves are emitted from the ship and travel to the seabed (sea floor).

2 Some of these waves are reflected from the seabed back up to the ship.

3 Equipment on the ship detects these sound waves.

4 The time it takes the waves to make this journey is measured.

5 Knowing this time, the depth of the sea below the ship can be calculated.

The system of using echoes in this way is called **sonar** (Sound, Navigation And Ranging).

ticking watch

cardboard tube

i

r

▲ Figure 13.4 Sound waves are reflected in the same way that light rays are reflected.

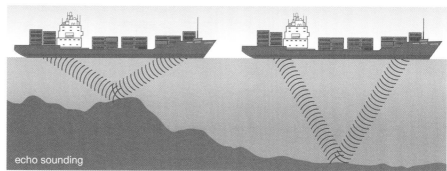

echo sounding

▲ Figure 13.5 Reflected sound can be used to tell ships about the depth of the sea beneath them.

REFRACTION OF SOUND

All waves can be refracted, even sound waves! For example, if some parts of a sound wave are travelling through warm air, they will travel more quickly than those parts travelling through cooler air. As a result the direction of the sound wave will change. It will be refracted.

Although it is not possible to see sound waves being refracted, we can sometimes hear their effect. Standing at the edge of a large pond or lake we can sometimes hear sounds from things on the other side of the water much more clearly than we would expect. This is due to refraction. Figure 13.6 explains how this happens.

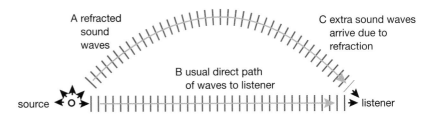

▲ Figure 13.6 Why sometimes sounds travelling across water are louder than we expect

1 Most of the sound we hear travels to us in a straight line (Path B).

2 But some sound travels upwards (Path A).

3 If the temperature conditions are right, then as the sound waves travel through the air they are refracted and follow a curved path downwards (Path C).

4 We now receive two sets of sound waves.

5 So the sound we hear seems louder and clearer.

PHYSICS ONLY

PITCH AND FREQUENCY

Small objects, such as the strings of the violin in Figure 13.7, vibrate quickly and produce sound waves with a high frequency. These sounds are heard as notes with a high pitch.

Larger objects, such as the strings of the cello, vibrate more slowly and produce waves with a lower frequency. These sounds have a lower pitch.

The frequency of a source is the number of complete vibrations it makes each second. We measure frequency in hertz (Hz). If a source has a frequency of 50 Hz this means that it vibrates 50 times each second and therefore produces 50 waves each second.

ACTIVITY 2

▼ PRACTICAL: INVESTIGATE THE FREQUENCY OF A SOUND WAVE USING AN OSCILLOSCOPE

Although we cannot see an actual sound wave, we can see an image or representation of it by connecting a microphone to a piece of apparatus called an **oscilloscope**. When the sound wave from a source such as a tuning fork or vibrating string enters the microphone, the oscilloscope 'draws' the longitudinal sound wave as a transverse wave which allows us to see features such as the wave's amplitude and frequency more easily.

▲ Figure 13.7 The violin produces notes that are higher pitched than those from the cello.

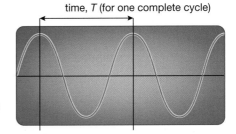

▶ Figure 13.8 An oscilloscope image of a sound wave

time, T (for one complete cycle)

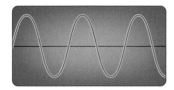 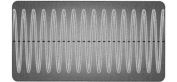

low-pitched sound   high-pitched sound

▲ Figure 13.9 Low-pitched sounds and high-pitched sounds seen on an oscilloscope. A low-pitched sound has a low frequency, meaning fewer complete waves per second, so fewer complete waves are seen on the oscilloscope than for the high-pitched (high frequency) sound.

From the trace drawn on the screen we can measure the time for one complete vibration or one complete wave. This is called the time period of the wave (T). We can then find the frequency of the sound (f) using the equation:

$$f = \frac{1}{T}$$

EXAMPLE 2

The picture of a sound wave seen on an oscilloscope has a time period of 0.005 s. What is the frequency of the wave?

$$f = \frac{1}{T}$$
$$= \frac{1}{0.005\ s}$$
$$= 200\ Hz$$

The relationship between the frequency (f), wavelength (λ) and speed of a wave (v) is described by the equation:

speed, v (m/s) = frequency, f (Hz) × wavelength, λ (m)

$$v = f \times \lambda$$

EXAMPLE 3

Calculate the wavelength of a sound wave that is produced by a source vibrating with a frequency of 85 Hz. The speed of sound in air is 340 m/s.

$$v = f \times \lambda$$
$$340\ m/s = 85\ Hz \times \lambda$$
$$\lambda = \frac{340\ m/s}{85\ Hz}$$
$$= 4\ m$$

The wavelength is 4 m.

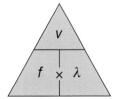

▲ Figure 13.10 The equation triangle for speed, frequency and wavelength.

AUDIBLE RANGE

The average person can only hear sounds that have a frequency higher than 20 Hz but lower than 20 000 Hz. This spread of frequencies is called the audible range or hearing range. The size of the audible range varies slightly from person to person and usually becomes narrower as we get older. You can demonstrate this range by using a signal generator and a loudspeaker to produce sounds at different frequencies.

You will not be tested on ultrasounds and infrasounds but it is interesting to know what they are.
Some objects vibrate at frequencies greater than 20 000 Hz. The sounds they produce cannot be heard by human beings and are called ultrasounds. Some objects vibrate so slowly that the sounds they produce cannot be heard by human beings. These are called infrasounds.

▲ Figure 13.11 Some animals can make and hear sounds that lie outside the human audible range. Dolphins can communicate using ultrasounds. Elephants can communicate using sounds that have frequencies too low for us to hear (infrasounds).

LOUDNESS

If the drum in Figure 13.12 is hit hard, lots of energy is transferred to it from the drum stick. The drum skin vibrates up and down with a large amplitude, creating sound waves with a large amplitude and we hear a loud sound. If the drum is hit more gently less energy is transferred and sound waves with a smaller amplitude are produced. We hear these as quieter sounds.

▲ Figure 13.12 If you hit a drum hard, you get a louder sound than if you beat it gently.

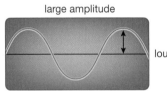
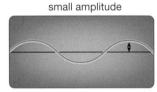

▲ Figure 13.13 Loud sounds and quiet sounds on an oscilloscope

▲ Figure 13.14 A scientist measuring the sound level as a car passes him

LOOKING AHEAD – SOUND WAVES IN DIFFERENT MATERIALS

Sound waves can travel through:

- solids – this is why you can hear someone talking in the next room, even when the door is closed
- liquids – this is why whales can communicate with each other when they are under water
- gases – the sound waves we create when we speak travel through gases (in the air).

Sound waves cannot travel through a vacuum because there are no particles to carry the vibrations.

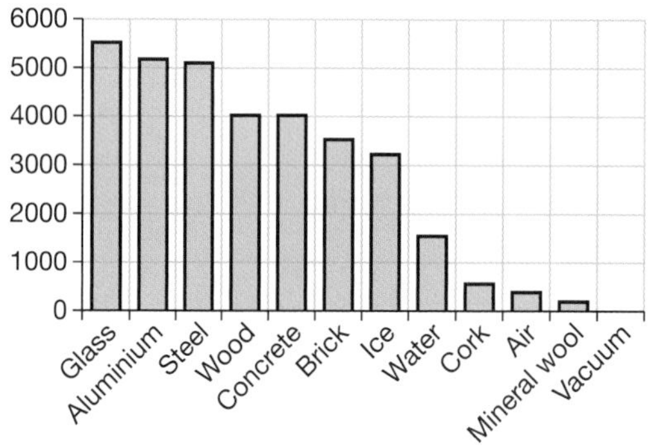

▶ Figure 13.15 Graph of speeds of sounds (in m/s) for different materials

CHAPTER QUESTIONS

More questions on sound and vibrations can be found at the end of Unit 3 on page 130.

SKILLS ▷ CRITICAL THINKING 4

1 a Name a musical instrument that is used to produce high-pitched notes.

SKILLS ▷ REASONING 5

  b Explain why the musical instrument you have named in part a produces high-pitched notes.

4

  c Explain how you would produce loud sounds from this musical instrument.

SKILLS ▷ INTERPRETATION 6

  d Draw the trace you might expect to see on an oscilloscope when this instrument is producing a loud, high-pitched note.

SKILLS ▷ CRITICAL THINKING 4

2 a What is an echo?

SKILLS ▷ REASONING 5

  b Explain how echoes are used by ships to find the depth of the ocean beneath them.

SKILLS ▷ PROBLEM SOLVING 7

  c A ship hears the echo from a sound wave 4 s after it has been emitted. If the speed of sound in water is 1500 m/s, calculate the depth of the water beneath the ship.

SKILLS ▷ CRITICAL THINKING 4

3 a What is meant by the phrase 'a person's audible range is 20 Hz to 20 000 Hz'?

5

  b Explain why the vibrating strings of a violin produce sounds with a higher frequency than those produced by the strings of a cello.

SKILLS ▷ PROBLEM SOLVING 8

  c Calculate the wavelength of sound waves whose frequency is 68 000 Hz. Assume that the waves are travelling through air at a speed of 340 m/s.

4 a An oscilloscope shows a wave which has a time period of 0.01 s. What is the frequency of this wave?

  b If the speed of this wave is 340 m/s, calculate its wavelength.

SKILLS ▷ CRITICAL THINKING, INTERPRETATION 7

5 a Sound waves are emitted from a source that is vibrating with a large amplitude and from a source that is vibrating with a small amplitude. Explain, using diagrams, the difference between the two sets of sound waves.

SKILLS ▷ INTERPRETATION 6

  b Draw two diagrams to show how these waves would appear on an oscilloscope.

END OF PHYSICS ONLY

UNIT QUESTIONS

SKILLS   CRITICAL THINKING   **1**

a Which of the following does not move as a transverse wave?

   A ultraviolet light

   B sound wave

   C surface water wave

   D microwave

                (1)

b Which of these effects describes the change in pitch we hear when a fast moving motorbike goes past?

   A total internal reflection

   B refraction

   C reflection

   D Doppler

                (1)

c Which of the following does not make use of total internal reflection?

   A oscilloscope

   B endoscope

   C prismatic periscope

   D optical fibres

                (1)

(Total for Question 1 = 3 marks)

SKILLS   INTERPRETATION   **2**   The diagram below shows the cross-section of a water wave.

a Copy this diagram and mark on it:

   i the wavelength of the wave (λ)   **(1)**

   ii the amplitude of the wave (A).   **(1)**

SKILLS   PROBLEM SOLVING   **b i** A water wave travelling at 20 m/s has a wavelength of 2.5 m. Calculate the frequency of the wave.   **(3)**

   ii Calculate the time period of the above wave.   **(1)**

(Total for Question 2 = 6 marks)

SKILLS   ANALYSIS   **3**   The diagram below shows a ray of light travelling down an optical fibre.

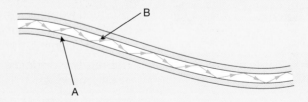

a What is A? **(1)**

b What is B? **(1)**

c Why is light reflected from the boundary between A and B? **(2)**

d Describe one medical use for optical fibres. **(1)**

(Total for Question 3 = 5 marks)

4

a i Explain the difference between a longitudinal wave and a transverse wave. **(1)**

ii Give one example of each type of wave. **(2)**

A girl stands 500 m from a tall building and bangs two pieces of wood together. At the same moment her friend starts a stopwatch. The sound waves created by the two pieces of wood hit the building and are reflected. When the two girls hear the echo they stop the stopwatch and note the time. The girls repeat the experiment four more times. The results are shown in the table below.

Experiment	Time (seconds)
1	2.95
2	3.00
3	2.90
4	3.20
5	2.95

b Why did the girls repeat the experiment five times? **(1)**

c Calculate the speed of sound using the results. **(6)**

d One of the girls thought that their answer might be affected by wind. Was she correct? Explain your answer. **(2)**

(Total for Question 4 = 12 marks)

5 The electromagnetic spectrum contains the following groups of waves: infrared, ultraviolet, x-rays, radio waves, microwaves, visible spectrum and gamma rays.

a Put these groups of waves in the order they appear in the electromagnetic spectrum starting with the group that has the longest wavelength. **(2)**

b Write down four properties that all of these waves have in common. **(4)**

c Write down one use for each group of waves. **(7)**

d Which three groups of waves could cause cancer? **(3)**

e Which three groups of waves can be used to communicate **(3)**

(Total for Question 5 = 19 marks)

6

a A ray of light hits the outside surface of a glass block with an angle of incidence of 38°. Its angle of refraction inside the block is 24°.

i Calculate the refractive index of the glass. **(4)**

ii Calculate the critical angle for this glass. **(2)**

b Calculate the refractive index for a piece of glass whose critical angle is 42°. **(3)**

(Total for Question 6 = 9 marks)

UNIT 4
ENERGY RESOURCES AND ENERGY TRANSFER

Energy, energy resources and the transfer of energy! These are all vital to modern life.

The photo shows energy produced from a wind farm. We also obtain energy by burning fossil fuels like coal and oil, directly from the Sun in the form of heat, from hydroelectric power and nuclear power, and from many other resources.

We need to understand how to transfer energy from one store to another, to use it efficiently and to conserve resources that cannot be replaced. We also need to be aware of the advantages and disadvantages of different energy resources.

Physics is about energy!

14 ENERGY TRANSFERS

Whenever anything happens, energy is transferred from one store to another – indeed, without energy things simply can't happen! In this chapter, you will learn that energy can be transferred to many different stores, including sound, light, movement, heat and potential energy. You will also find out that, although energy is never destroyed, in every energy transfer some energy is transferred to the surroundings, often as heat.

▲ Figure 14.1 Energy is stored as sound, movement and light, for example.

LEARNING OBJECTIVES

■ Describe energy transfers involving energy stores:

  ■ energy stores: chemical, kinetic, gravitational, elastic, thermal, magnetic, electrostatic, nuclear

  ■ energy transfers: mechanically, electrically, by heating, by radiation (light and sound)

■ Use the principle of conservation of energy

■ Know and use the relationship between efficiency, useful energy output and total energy output:

$$\text{efficiency} = \frac{\text{useful energy output}}{\text{total energy output}} \times 100\%$$

■ Describe a variety of everyday and scientific devices and situations, explaining the transfer of the input energy in terms of the above relationship, including their representation by Sankey diagrams

UNITS

In this section you will need to use kilogram (kg) as the unit of mass, joule (J) as the unit of energy, metre (m) as the unit of length, metre per second (m/s) as the unit of speed and velocity, metre per second squared (m/s^2) as the unit of acceleration, newton (N) as the unit of force, second (s) as the unit of time and watt (W) as the unit of power. You have used all these units before.

For things to happen we need energy! Energy is used to produce sound. Energy is used to transport people and goods from place to place, whether it is by train, boat or plane or on the backs of animals or even by bicycles. Energy is needed to lift objects, make machinery work and run all the electrical and electronic equipment we have in our modern world. Energy is needed to make light and heat. The demand for energy increases every day because the world's population is increasing. People consume energy in the form of food and need energy for the basics of life, like warmth and light. As people become wealthier they demand much more than the basics, so the need for energy grows!

STORES OF ENERGY

Energy is found in many different stores:

■ We get our energy from the food we eat. Food provides stored chemical energy that we can burn to transfer it to other types of energy.

■ We use the energy from food to generate **thermal** energy (heat energy) to help to keep us warm.

■ Our muscles transfer the chemical energy to movement energy (kinetic energy).

■ Some of this movement energy is transferred as we speak to sound energy.

■ We need heat for our homes, schools and workplaces. We also need to transfer energy to light for our buildings, vehicles and roads.

■ Most of the energy needed for these purposes is transferred from electrical energy. We shall see later (see Chapter 17) that electrical energy can be transferred from other stores of energy, like chemical or nuclear energy.

■ Some electrical energy is produced from the gravitational potential energy in water kept in reservoirs in mountainous areas. We can also use the energy in the hot core (centre) of the Earth. We see the evidence of this huge supply of heat in volcanoes and thermal springs.

■ Energy from the heat underground is called geothermal energy.

■ Energy can also be stored in springs as elastic potential energy. This type of stored energy is used in things like clocks and toys.

The main resource of energy for the Earth is our Sun. This provides us with heat, light and other stores of energy.

ENERGY TRANSFERS

For energy to be useful, we need to be able to transfer it from one store into whichever store we require. Unfortunately, when we try to do this there is usually some energy transferred to unwanted stores. We often refer to these unwanted stores as 'wasted' energy because it is not being used for a useful purpose.

Here are some examples of wasted energy:

■ An electric heater may be used to heat water in a house. The hot water will be stored in a tank (container) for use when required. Although the tank may be well insulated, some energy will be transferred from the water and some will be used to heat up the copper that the tank is made from. Both processes mean that some of the energy that is transferred from electricity to heat is wasted because it is not doing what we want it to do – it is not making the water hot.

■ When we fill a car with petrol, the main purpose is to transfer the chemical energy stored in the fuel to movement energy, with a small amount doing other things like providing electrical energy for the lights or radio. But the process of transferring the energy is not perfectly efficient, as not all of the energy is used to do what we want. (A formal definition of efficiency is given on page 137.) A considerable amount of the energy supplied by the fuel is transferred to heat, most of which is transferred to the surroundings. Some of the energy is transferred to sound, which can be unpleasant for people both inside and outside the car. A lot of energy is transferred to overcome the various friction forces that oppose movement of parts within the car and the car's movement along the road. Friction causes energy transfers to occur too, usually producing unwanted heat energy.

Unwanted energy transfers reduce efficiency. This problem is the same whether the system is a small one, like a car, or a large system, like the nationwide electricity generation and distribution industry. We need to be aware of where our energy is transferring to if we are to find ways of using it well.

We have many ways of transferring energy from one store to another.

  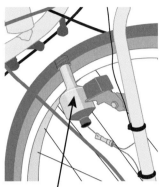 

dynamo

▲ Figure 14.2 Energy is transferred from one store to another, and to another, and so on.

In Figure 14.2, stored chemical energy in the food is needed to help our bodies make a range of other chemicals. Some of these – like carbohydrates – are used to produce heat to maintain our body temperature and energy for movement through muscle activity. Having eaten a meal, the cyclist in Figure 14.2 is transferring the chemical energy stored in his body to movement energy. The movement is initially in the cyclist's legs and is then transferred to the machine (the bicycle). The cyclist is also transferring additional heat energy, which is then transferred to the surroundings. Friction in various parts of the bicycle will also result in energy being transferred from movement to heat and sound.

The dynamo or generator fitted to the bicycle's wheel transfers some of the movement energy of the wheel to electrical energy. The lamp then transfers the electrical energy to light and heat. As the electrons flow in the dynamo–lamp circuit, heat will be produced in the conducting wires too.

Examples of other energy transfers are given in the questions at the end of the chapter.

CONSERVATION OF ENERGY

The principle of conservation of energy is a very important rule. It states that:

Energy is not created or destroyed in any process.

(It is just transferred from one store to another.)

We often hear about the energy crisis: as our demand for more energy increases our reserves of energy in the form of fuels like oil and gas are rapidly being used up. The principle of conservation of energy makes it seem as if there is no real problem – that energy can never run out. We need to understand what the principle really means.

Physicists believe that the amount of energy in the Universe is constant – energy can be transferred from one store to another but there is never any more or any less of it. This means we cannot use energy up. However, if we consider our little piece of the Universe, the problem becomes more obvious. As we make energy do useful things – for example, move a car – some of it will be transferred to heat. Some of this heat energy will be radiated away from the Earth and transferred (not destroyed or used up) into space. This means the

Even though the amount of energy in the Universe is constant, it is becoming more spread out and so less available for use. Some scientists think that all the energy in the Universe will eventually be transferred to heat and that everywhere in the Universe will end up at the same very low temperature. This possible 'end of the world' is sometimes referred to by the dramatic name of 'heat death'. If the Universe does end up this way, it is not expected to do so for some time yet, so carry on working for those exams!

energy is not available for us to use any more. This is just like a badly insulated house; if heat energy escapes, it is transferred away from the system we call our home, and is no longer available to keep us warm.

SANKEY DIAGRAMS

We use different ways to show how energy is transferred. Energy transfer diagrams show the energy input (contribution), the energy transfer process and the energy output (production). The system may be a very simple one with just one main energy transfer process taking place. An example of a simple system with its energy transfer diagram is shown in Figure 14.3.

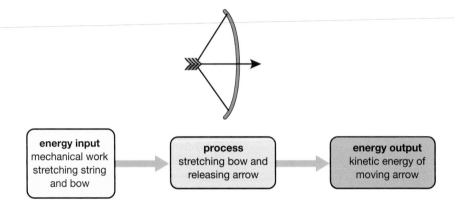

▲ Figure 14.3 Energy transfer diagram for an arrow being fired from a bow

Sankey diagrams are a simpler and clearer way of showing what happens to an energy input into a system. The energy flow is shown by arrows whose width is proportional to the amount of energy involved. Wide arrows show large energy flows, narrow arrows show small energy flows.

Figure 14.4 shows a Sankey diagram for a complex system – the energy flow for a car. Chemical energy in the form of petrol is the input to the car. The energy outputs from the car are:

■ electrical energy (from the alternator) to drive lights, radio and so on, to charge the battery (transferred to chemical energy) and allow the car to switch on

■ movement (kinetic) energy from the car engine

■ wasted energy as electrical heating in wiring and lamp filaments, as frictional heating in various parts of the engine and alternator, and as noise.

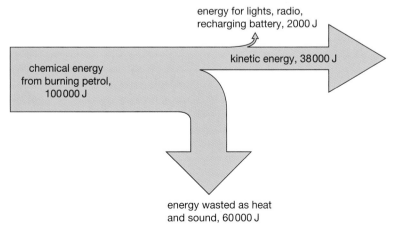

▲ Figure 14.4 Sankey diagram showing the energy flow in a typical car

Here the 100 000 J of chemical energy input might be shown by an arrow 20 mm wide, so the 60 000 J of energy wasted (60% of the input) would then be shown by an arrow that is 12 mm wide (60% of 20 mm). It is difficult to draw the 2% arrow for the energy output to scale; it is enough to show it as very small.

EFFICIENCY

Whenever we are considering energy transfers, we have to remember that a proportion of the energy input is wasted. Remember that wasted means transferred into stores other than the useful store required. We would like our energy transfer systems to be perfect with all the output energy being in the store that we want. For example, not all the output energy for an electric lamp is light (the useful energy output), some of the output energy is heat (not useful when what we want is light).

Real systems always have an unwanted energy output so can never have 100% efficiency.

The efficiency of an energy conversion system is defined as:

$$\text{efficiency} = \frac{\text{useful energy output}}{\text{total energy output}} \times 100\%$$

Efficiency does not have a unit because it is a ratio. Sometimes efficiency is shown as a **fraction** of the energy output that is in the wanted or useful store. In real energy transfers this fraction will always be less than 1 because some of the total output energy will be in an unwanted store.

EXAMPLE 1

A 60 W tungsten filament bulb uses 60 J of energy every second. It is 5% efficient. How much of the total energy output per second is useful light energy?

$$\text{efficiency} = \frac{\text{useful energy output}}{\text{total energy output}} \times 100\%$$

$$5\% = \frac{\text{useful energy output}}{60 \text{ J}} \times 100\%$$

$$\frac{5}{100} = \frac{\text{useful energy output}}{60 \text{ J}}$$

So useful (light) energy from bulb $= \dfrac{5 \times 60 \text{ J}}{100} = 3$ J each second.

CHAPTER QUESTIONS

More questions on the need for energy can be found at the end of Unit 4 on page 168.

SKILLS CREATIVITY

1 Describe the main energy transfers taking place in the following situations:

  a turning on a torch

  b lighting a candle

  c rubbing your hands to keep them warm

  d bouncing on a trampoline.

2 Copy and complete the following Sankey diagrams. Remember that the width of the arrows must be proportional to the amount of energy involved. This has been done for you in part a.

a for an electric lamp

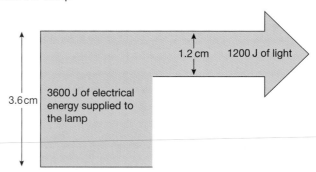

3.6 cm

3600 J of electrical energy supplied to the lamp

1.2 cm   1200 J of light

b In a typical wash in a washing machine 1.2 MJ is transferred to the kinetic energy of the rotating drum, 6 MJ of energy is transferred to heat the water and 0.8 MJ is wasted as heat transferred to the surroundings and sound.

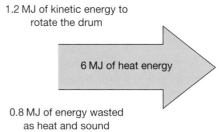

1.2 MJ of kinetic energy to rotate the drum

6 MJ of heat energy

0.8 MJ of energy wasted as heat and sound

3 a Draw a Sankey diagram for the following situation. An electric kettle is used to heat some water. 350 kJ of energy are used to heat the water, 10 kJ raise the temperature of the kettle and 40 kJ escape to heat the surroundings.

b Calculate the efficiency of the kettle.

4 A ball is dropped. It hits the ground with 10 J of kinetic energy and bounces with 4 J of kinetic energy.

a What happens to 6 J of the energy during the bounce?

b Draw a Sankey diagram for the energy flow that takes place during the bounce.

15 THERMAL ENERGY

This chapter is about the ways in which thermal (heat) energy is transferred from a hotter place to a cooler place.

▲ Figure 15.1 The transfer of thermal energy from place to place can be useful – for example, rising warm air creates 'thermals' that can carry this glider up to great heights.

LEARNING OBJECTIVES

- Describe how thermal energy transfer may take place by conduction, convection and radiation

- Explain the role of convection in everyday phenomena

- Explain how emission and absorption of radiation are related to surface and temperature

- Practical: investigate thermal energy transfer by conduction, convection and radiation

- Explain ways of reducing unwanted energy transfer, such as insulation

Thermal or heat energy is energy that is stored in 'hot' matter. We shall see that 'hot' is a relative term. There is a temperature called absolute zero that is the lowest possible temperature. Any matter that is above this temperature has some thermal energy. The kinetic energy of the minute particles that make up all matter produces the effect we call heat.

Thermal energy is transferred from a place that is hotter (that is, at a higher temperature) to one that is colder (at a lower temperature). In this chapter, we will look at the different ways in which thermal energy is transferred between places that have different temperatures.

CONDUCTION

Thermal **conduction** is the transfer of thermal (heat) energy through a substance by the vibration of the atoms within the substance. The substance itself does not move.

If you have ever cooked kebabs on a barbecue with metal skewers, as shown in Figure 15.2, you will have discovered conduction! The metal over the burning charcoal becomes hot and the heat energy is transferred along the skewer by conduction. In metals, this takes place quite rapidly and soon the handle end is almost as hot as the end over the fire. Metals are good thermal

▲ Figure 15.2 Metal skewers allow heat to be transferred to parts that are away from the heat.

> **!** The metal rods will quickly get hot enough to burn the skin and will remain hot long after the heat is removed.

conductors. If you use skewers with wooden handles you can hold the wooden ends much more comfortably because wood does not conduct thermal energy very well. Wood is an example of a good thermal insulator.

The process of energy transfer by conduction is explained in terms of the behaviour of the tiny particles that make up all matter. In a hot part of a substance, like the part of the skewer over the hot charcoal, these particles have more kinetic energy. The more energetic particles transfer some of their energy to particles near to them. These therefore gain energy and then pass energy on to particles near to them. The energy transfer goes on throughout the substance. This process takes place in all materials.

In metals, the process takes place much more rapidly, because metals have free electrons that can move easily through the structure of the metal, making the transfer of energy happen faster.

ACTIVITY 1

▼ PRACTICAL: INVESTIGATE HOW WELL DIFFERENT METALS CONDUCT HEAT

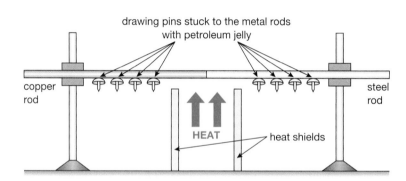

▲ Figure 15.3 Experiment to show thermal conduction in different metals

Figure 15.3 demonstrates how heat is transferred along the metal rods from the heated ends towards the cooler ends away from the heat source. As the heat passes along the rods the petroleum jelly holding the drawing pins in place melts and they drop off in sequence. This experiment also shows that copper conducts heat better than steel because the drawing pins attached to the copper rod drop off sooner than those attached to the steel rod. The rods should be the same diameter and the drawing pins placed at the same distances from the heat source for this to be a fair test.

CONVECTION

Convection is the transfer of heat through fluids (liquids and gases) by the upward movement of warmer, less dense regions of fluid.

You may have seen a demonstration of convection currents in water, like the one shown in Figure 15.4. The water is heated just under the purple crystal and the crystal colours the water as it **dissolves**, which lets you see the movement in the water. The heated water expands and becomes less dense than the colder surrounding water, so it floats up to the top of the glass beaker. Colder water sinks to take its place, and is then heated too. At the top, the warm water starts to cool, becomes more dense again and will begin to sink, so a circulating current is set up in the water. This is called a convection current.

ACTIVITY 2

▼ PRACTICAL: INVESTIGATE CONVECTION CURRENTS IN WATER

> ⚠ Avoid skin contact with the crystals and their solution.

▲ Figure 15.4 Demonstration of convection currents, using a potassium manganate (VII) crystal in water

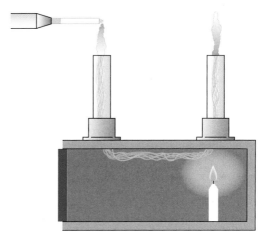

▲ Figure 15.5 Here, air warmed by the candle floats up the chimney on the right and colder air is drawn down the chimney on the left. Smoke from the burning candle shows up the circulation of the air.

EXTENSION WORK

Substances tend to expand when heated because the particles of which they are made have more kinetic energy. As they move around more, the average distance between the particles increases.

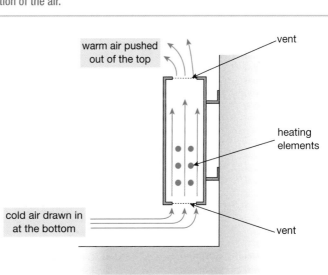

warm air pushed out of the top

vent

heating elements

cold air drawn in at the bottom

vent

▲ Figure 15.6 A convector heater relies on the effects of convection.

Convection occurs in any fluid substance – that is, in things that can flow, such as liquids and gases. Convector heaters (Figure 15.6) heat air, which then floats out of the top of the heater to the top of the room. Cold air is drawn in at the bottom and this in turn is heated. In this way, heat energy is eventually transferred to all parts of the room.

In many cooking ovens, the heating element is placed at the bottom of the oven. It heats the air near to it, and this air rises by convection. The top of the oven is generally warmer than the bottom, so you can cook foods at different temperatures. However many modern ovens are fan ovens, where hot air is blown into the oven and provides an even temperature throughout the oven.

Air and water both allow heat transfer to take place by convection as they are both fluids, but neither are good thermal conductors (they are insulators). This insulating property of both water and air is put to good use in situations where they are not able to circulate easily. For example, woollen clothing keeps you warm because air gets trapped in the fibres. The trapped air is heated by your body and forms a warm insulating layer that helps to stop you losing heat. In the same way, a wetsuit keeps a diver warm because a thin layer of water is trapped next to the diver's skin. (Figure 15.7)

▲ Figure 15.7 A diver wearing a wetsuit to keep warm in cold water

Convection currents are responsible for many everyday events. One example is on-shore and off-shore breezes, also known as sea breezes and land breezes. These are explained in Figure 15.8.

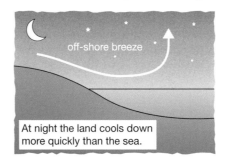
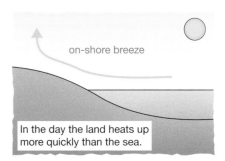

▲ Figure 15.8 At night the air over the warmer sea floats up causing cooler air to flow towards sea. During the day the situation is reversed. In Unit 5 you will learn that some things heat up and cool down more easily than others.

Another example is convection currents within the very tall 'thunder' clouds that are responsible for the build-up of charge at the cloud base resulting in lightning strikes.

RADIATION

Thermal radiation is the transfer of energy by infrared (IR) waves.

REMINDER

Remember that we are talking about heat transfer – this is thermal radiation not nuclear radiation!

▲ Figure 15.9 Heat is transferred from a heater by radiation.

The reflector in an electric fire is a special shape, called a parabola. You can see parabolic reflectors in torches and radio telescopes, for example.

When you turn on a bathroom heater, as shown in Figure 15.9, you will feel the effect almost instantly. Neither conduction nor convection can explain how heat is getting from the hot part to your hands. Conduction does not occur that rapidly, even in good thermal conductors, and air is a poor thermal conductor. Convection results in heated air floating upwards on colder, denser air.

There are two things you should notice about this example.

1 The heat that you feel so quickly is travelling from the heater in a straight line.

2 The design of the bathroom heater includes a specially shaped, very shiny reflector, similar to the reflector behind a fluorescent light or in a torch.

In this example, heat is travelling in the form of waves, like visible light. Heat waves are called infrared (IR) waves or IR radiation. The army and the emergency services use special cameras, called thermal imaging cameras, that can detect objects giving out IR waves. These cameras show images of people because of the heat radiation from their bodies, even when there is not enough visible light to actually see them. Thermal imaging is also an important tool in the diagnosis of certain illnesses. (Figure 15.10)

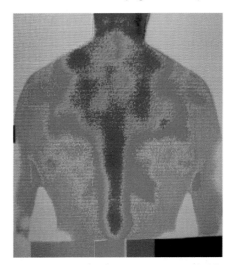

▲ Figure 15.10 This is a thermal image of a patient showing areas of different temperatures.

IR waves are part of the same family of waves as light, radio waves, ultraviolet and so on, called the electromagnetic (EM) spectrum (see page 106). IR waves, therefore, have the same properties as all the other waves in the EM spectrum. In particular, IR can travel through a vacuum and does so at the speed of light $(3 \times 10^8 \text{ m/s})$.

It is important that heat can travel in this way, without the need for matter, otherwise we would not receive heat, as well as light, from the Sun.

IR waves can also be reflected and absorbed by different materials, just like visible light. Highly polished, shiny surfaces are good reflectors of thermal radiation. White surfaces also reflect a lot of IR. Matt (not shiny) black and dark surfaces are poor reflectors or, to put it more positively, are good absorbers of heat radiation. Figure 15.12 shows how this can be useful in everyday life.

ACTIVITY 3

▼ PRACTICAL: INVESTIGATE HOW WELL DIFFERENT SURFACES RADIATE HEAT

This experiment shows that matt black surfaces radiate heat better than shiny white surfaces.

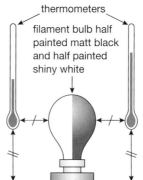

thermometers

filament bulb half painted matt black and half painted shiny white

▲ Figure 15.11 Demonstrating that matt black surfaces radiate heat better than shiny white surfaces

Figure 15.11 shows the experiment. Put two identical (same type) thermometers on either side of a filament bulb that has been painted matt black on one side and shiny white on the other.

When you turn on the bulb you will notice that the temperature starts to rise more quickly on the thermometer facing the black side than on the other.

It is important that the thermometers are fixed at the same height and distance from the filament bulb.

The bulb will quickly get hot enough to burn skin and ignite paper.

▲ Figure 15.12 Shiny and white surfaces reflect thermal radiation, while matt black surfaces, like in the solar heating panels, absorb it.

a   b

▲ Figure 15.13 **a** A shiny kettle stays warmer longer. **b** The heat sink needs to be matt black to lose heat to the surroundings quickly, and so stop the transistor overheating.

If a surface is a good reflector of IR then it is a poor radiator of IR. This means that a hot object with a shiny surface will emit less heat energy in the form of IR than another object at the same temperature with a matt black surface. The kettle in Figure 15.13a has a shiny surface to reduce the rate of heat loss. Heat sinks are used in electronic equipment to stop parts getting too hot. A heat sink is shown in Figure 15.13b. The transistor is fixed to the black metal heat sink and heat is transferred from the transistor to the heat sink by conduction. The matt black surface radiates heat well and the shape of the heat sink helps convection air currents to transfer heat away from the heat sink.

The amount and type of the energy radiated by a hot object does not only depend on its surface texture (how it feels to the touch) and colour – it also depends on how hot it is. Not only does the amount of energy radiated per second increase significantly with temperature but the nature of the EM waves also changes. At a low temperature most of the radiated energy is in the form of infrared waves (invisible to the human eye). As the temperature of a metal object increases it starts to radiate in the visible spectrum as well. Things that do not burn will start to glow a dull red. As the temperature rises further the colour changes through the visible spectrum, for example, the tungsten filament glows white hot when it reaches ~3000 °C.

You will meet this effect again in Unit 8 Astrophysics.

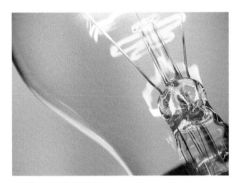

▲ Figure 15.14 A tungsten filament bulb

ENERGY-EFFICIENT HOUSES

We pay for the energy we use in our homes, schools and places of work. Heating is the main use of energy in our homes and – since most domestic heating systems work by burning fuels like coal, oil and gas – it is the main producer of carbon dioxide. (Even if electric heaters are used, most electrical energy is produced by burning fuels in power stations.) Carbon dioxide is a greenhouse gas and contributes to global warming. It is, therefore, very important that houses are energy efficient.

Energy efficiency means using as much as possible of the energy we produce for the desired purpose. So when we turn on the central heating, we want to keep the insides of our homes warm and not allow the heat to escape. If no heat can escape from a house then we will only need to heat it until it reaches the desired temperature.

The key to energy-efficient housing is insulation. Houses must be designed to reduce the rate at which energy is transferred between the inside and the outside.

To insulate a house effectively we must look at all the ways in which heat energy can escape. Conduction is the main way heat is transferred between the inside of a building and the outside. Next we need to consider the places where conduction occurs: the walls, the windows (and doors) and the roof.

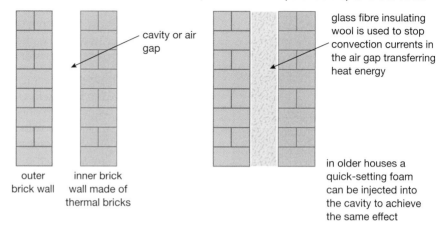

cavity or air gap

glass fibre insulating wool is used to stop convection currents in the air gap transferring heat energy

outer brick wall

inner brick wall made of thermal bricks

in older houses a quick-setting foam can be injected into the cavity to achieve the same effect

▲ Figure 15.15 Two-layered wall construction, with the gap filled with insulation panels, helps to reduce heat loss by conduction, convection and even radiation.

Heat loss by conduction through the walls can be reduced by using building materials that are good insulators. However, the materials used for building must also have other suitable properties like strength, durability and availability at a sensible price. For walls, bricks are a common building material.

Figure 15.15 shows the typical construction of a modern house in the UK built to follow the current energy efficiency regulations.

As you can see, the wall is made with layers of different materials. The outer layer is made with bricks – these have quite good insulating properties, are strong and will survive bad weather conditions. The inner layer is built with thermal bricks with very good insulation properties – they are also light, relatively cheap and quick to work with. The two layers of brick are separated by an excellent thermal insulator in the form of an air cavity or gap.

The walls also stop heat being lost by convection. The cavity or gap between the two walls is wide enough for convection currents to circulate. This means heat is circulated from the warmer surface of one wall to the colder surface of the other. To stop convection currents, the gap in modern houses is filled with insulating panels made of glass fibre matting. This is a lightweight, poor conductor that traps lots of air. The panels are usually surfaced with thin aluminium foil. This highly reflective surface reflects heat in the form of infrared radiation.

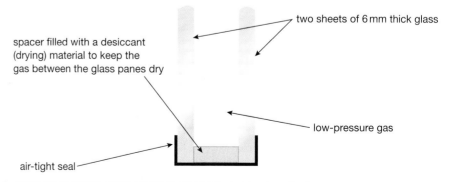

two sheets of 6 mm thick glass

spacer filled with a desiccant (drying) material to keep the gas between the glass panes dry

low-pressure gas

air-tight seal

▲ Figure 15.16 Double glazing helps to stop heat escaping from the home.

Figure 15.16 shows a cross-section of a typical double-glazed window, as used in modern houses. Glass is a poor thermal conductor but is used in thin layers. To improve the insulating properties, two layers of glass are used to trap a layer of air. The thickness of this layer is important. If it is too thin then the insulation effect is reduced, but if it is too thick then convection currents will be able to circulate and carry heat from the hotter surface to the colder one. In very cold countries triple glazing is used. Modern double glazing uses special glass to increase the greenhouse effect (heat radiation from the Sun can get in but radiation from inside the house is mainly reflected back again).

Roof insulation in modern houses uses similar panels to those used in the wall cavities, trapping a thick layer of air. This takes advantage of the poor conducting property of air, whilst also preventing convection currents circulating. Again, reflective foil is used to reduce radiation heat loss. Figure 15.17 shows houses built before loft insulation was a compulsory building regulation. In some, the owners have installed loft insulation – you should be able to identify which!

There are other things that can be done to improve the energy efficiency of houses that do not relate directly to the mechanisms of heat transfer discussed in this chapter. For example, thermostats and computer control systems for central heating can further reduce the heating needs of a house. They stop rooms being heated too much by switching off the heat when a certain temperature is reached. Another important energy-saving measure is the reduction or elimination of draughts (air currents) from poorly fitting doors and windows.

With your understanding of how heat travels you can save your family money, keep warm and reduce global warming.

▲ Figure 15.17 Loft insulation makes a big difference to heat loss through the roof.

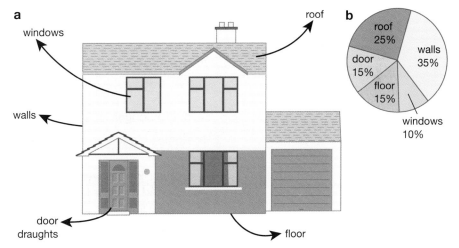

▲ Figure 15.18 **a** How heat energy can be lost from the home; **b** Percentage of energy lost in different ways

INSULATING PEOPLE AND ANIMALS

Earlier in this chapter we saw a picture of a fire fighter in protective clothing designed to reduce the amount of heat getting to their bodies (Figure 15.12). Sometimes we have the opposite problem and want to keep warm. The obvious method of cutting down heat loss from the body is to wear clothes. Clothes that trap air around the body provide insulation because trapped air cannot circulate and is a very poor conductor. A large proportion of body heat is lost from the head, so hats are the human equivalent of loft insulation.

Wind can cause rapid heat loss from the body. It does this by forced convection – that is, making air circulate close to the body surface. It may also cause sweat to evaporate from the skin more quickly, causing rapid cooling. (The purpose of sweat is to help the body to lose heat by evaporation, but, if it happens because of strong wind on a cold day, the effect can be life threatening.) These cooling effects of wind contribute to what is called the wind-chill factor. To reduce the wind-chill effect, a piece of wind-proof outer clothing should be worn.

When people do lose body heat at too great a rate they may become hypothermic, which means their body temperature starts to fall. If the heat loss is not significantly reduced the condition is potentially fatal. When people are rescued from mountains suffering from the effects of cold they are usually wrapped in thin, highly reflective blankets. The interior reflective surface reflects heat back to their bodies while the outer reflective surface is a poor radiator of heat. Marathon runners are often covered in these blankets at the end of the race to keep them warm when their energy reserves are low.

▲ Figure 15.19 Penguins stand close together for warmth.

Animals keep warm in different ways. You may have noticed birds fluffing up their feathers on cold days in winter. This increases the thickness of the trapped air layer around their bodies, so reducing heat loss by conduction. Some birds, like penguins, will move close together for warmth (Figure 15.19). Other animals will curl into small balls. This cuts down heat loss by making the surface area of their bodies exposed to the cold as small as possible.

CHAPTER QUESTIONS

SKILLS CREATIVITY, REASONING

More questions on thermal energy can be found at the end of Unit 4 on page 168.

1 Explain the following observations, referring to the appropriate process of heat transfer in each case.

 a Two cups of tea are poured at the same time. They are left for ten minutes. One of the cups has a metal teaspoon left in it. The tea in this cup is cooler than the tea in the other cup at the end of the ten-minute period.

 b Two fresh cups of tea are poured. (The others had gone cold!) A thin plastic lid is placed on top of one of the cups. The tea in this cup keeps hot for longer.

SKILLS REASONING

2 a Kettles heated on stoves used to be made of copper. Was this a good choice?

 b Copper kettles were usually kept highly polished (shiny). If it is not polished, copper turns matt and eventually blackens as it reacts with oxygen in the air. Apart from making the kettle look nice, what is a good physics reason for keeping a copper kettle polished?

3 The diagrams below show a physics demonstration about thermal conduction.

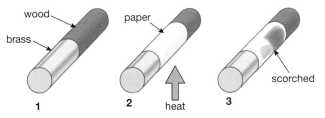

A cylinder is made from a piece of brass fitted to a piece of wood. A piece of paper is glued around the middle. The paper is then heated over a Bunsen burner flame. After a while one end of the paper is noticeably more burnt than the other. Explain why this happens.

4 There are two bench seats in a park, one made of metal, the other made of wood. The metal seat feels much colder to sit on than the wooden one. A student says that it is because the metal seat is at a lower temperature than the wooden one. Explain why this explanation is incorrect, and give a correct explanation of why the metal seat seems colder than the wooden one.

5 a Why is the heating element (part) in an electric kettle positioned very close to the bottom of the kettle?

b Where would you expect the cooling element to be placed in a freezer? Give a reason for your answer.

6 One model of a well-known brand of computer does not use a fan to keep the electronic circuits inside it cool, unlike other PCs. A student noticed that the ventilation (air) slots on most other PCs are positioned on the side, but the slots are on the top and bottom surfaces of this computer. The designer has applied physics to the problem of keeping the computer cool. Explain why the new computer does not need a fan.

7 The diagram below shows how Roman mines used to be ventilated.

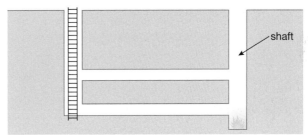

The mine system had a shaft with a fire lit at the bottom. Explain how this kept the air in the mine system fresh.

8 a Describe an experiment to show that matt black surfaces absorb thermal radiation better than shiny metallic surfaces. Your description should include details of the apparatus; you may use a clearly labelled sketch diagram in your answer.

b State what measurements you would take.

c Explain how you would make sure your investigation is fair.

d How would your results show that the idea that a matt black surface absorbs thermal radiation better than a shiny metallic surface is correct?

16 WORK AND POWER

Work is calculated by multiplying the force applied by the distance through which the force moves – the bigger the force, or the longer the distance through which it moves, the more work is done. Work always involves an energy transfer. Power is the rate at which energy is transferred, and efficiency is a measure of how much of the input energy to a system is converted to useful output energy. In this chapter you will learn how to calculate the work done in a system and its power as energy is transferred.

▶ Figure 16.1 James Joule (1818–1889) was the son of a wealthy Manchester brewer. He was tutored by James Dalton and carried out scientific research in his own laboratory, built in the basement of his father's home.

LEARNING OBJECTIVES

- Know and use the relationship between work done, force and distance moved in the direction of the force:

  work done = force × distance moved

  $W = F \times d$

- Know that work done is equal to energy transferred

- Know and use the relationship between gravitational potential energy, mass, gravitational field strength and height:

  gravitational potential energy = mass × gravitational field strength × height

  $GPE = m \times g \times h$

- Know and use the relationship:

  kinetic energy = $\frac{1}{2}$ × mass × speed squared

  $KE = \frac{1}{2} \times m \times v^2$

- Understand how conservation of energy produces a link between gravitational potential energy, kinetic energy and work

- Describe power as the rate of transfer of energy or the rate of doing work

- Use the relationship between power, work done (energy transferred) and time taken:

  power = $\frac{\text{work done}}{\text{time taken}}$

  $P = \frac{W}{t}$

The unit of energy is named after James Joule. It was Joule who realised that heat was a store of energy. He showed that kinetic energy could be transferred to heat. At that time heat was measured in calories.

ENERGY AND WORK

Energy is the ability to do work.

This statement tells us what energy does rather than what energy is. We know that energy is found in a wide variety of different stores but we are really interested in what energy can do – the answer is that energy does work.

We need to define work in a way that is measurable. Some types of work are not easy to calculate the value of. Mechanical work, like lifting heavy objects, is easy to measure: if you lift a heavier object, you do more work; if you lift an object through a greater distance, again, you do more work. The definition of work in physics is:

work done, W (joules) = force, F (newtons) × distance moved, d (metres)

$$W = F \times d$$

If the force is measured in newtons and the distance through which the force is applied is measured in metres then the work done will be in joules.

Work done is equal to the amount of energy transferred.

1 J of work done is transferred when a force of 1 N is applied through a distance of 1 m in the direction of the force.

EXAMPLE 1

Figure 16.2 shows a weightlifter raising an object that weighs 500 N through a distance of 2 m. To calculate the work done we use:

$W = F \times d$

$\quad = 500\ \text{N} \times 2\ \text{m}$

$\quad = 1000\ \text{J}$

height lifted 2 m

weight 500 N

▲ Figure 16.2 Doing work by lifting a weight

This work done on the weight has increased its energy. This is explained in the section on gravitational potential energy (page 152).

EXAMPLE 2

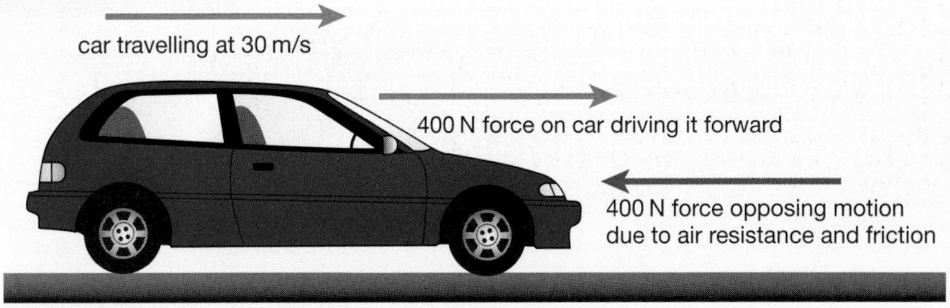

car travelling at 30 m/s

400 N force on car driving it forward

400 N force opposing motion due to air resistance and friction

▲ Figure 16.3 A car travelling at a constant speed doing work

In the example shown in Figure 16.3 the force acting on the car is not accelerating it – instead, it is being used to balance the forces opposing its movement. The resultant force on the car is zero, so it keeps moving in a straight line at constant speed. To work out the work done on the car in one second we substitute the force required, 400 N, and the distance through which it acts in one second, 30 m, in the equation:

$W = F \times d$

$\quad = 400\ \text{N} \times 30\ \text{m}$

$\quad = 12\,000\ \text{J or 12 kJ}$

GRAVITATIONAL POTENTIAL ENERGY (GPE)

The gravitational potential energy of an object that has been raised to a height, h, above the ground is given by:

gravitational potential energy, GPE (joules) = mass of object, m (kilograms)

× gravitational field strength, g (newtons per kilogram) × height, h (metres)

$$GPE = m \times g \times h$$

The change in the GPE of an object will be an increase if we apply a force on it in the opposite direction to the pull of gravity – that is, if we lift it off the ground. When an object falls it loses GPE. To keep things simple, we usually assume that an object has no GPE before we do work on it.

In the weight-lifting example given on page 151, the weightlifter has used some chemical energy to do the work. We know that energy is conserved so what has happened to the chemical energy that the weightlifter used? Some has been transferred to heat in the weightlifter's body. The remainder has been transferred to the weight because he has increased its height in the gravitational field of the Earth. The energy that the weight has gained is called gravitational potential energy or GPE.

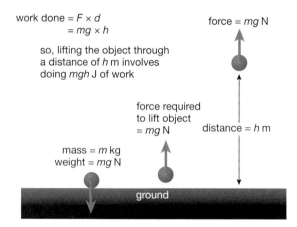

▲ Figure 16.4 The work done to lift an object is equal to the GPE the object has at its new height.

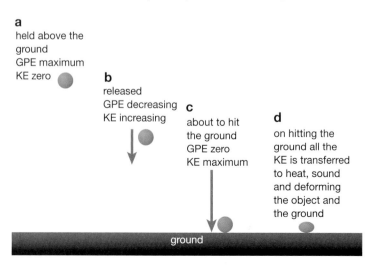

▲ Figure 16.5 When a raised object falls, its gravitational potential energy is transferred first to kinetic energy and then to heat and sound.

Kinetic energy (KE) is the energy stored by moving objects.

In Figure 16.5, we can see the GPE stored by the weight is being transferred to other stores as the weight falls. The weight accelerates because of the force of gravity acting on it, so it gains kinetic energy. When it reaches the ground all the initial GPE is transferred to kinetic energy. When it hits the ground all the movement energy is then transferred to other stores, mainly heat and sound.

REMINDER

Gravitational field strength is the force acting per kilogram on a mass in a gravitational field. The gravitational field strength, g, on the surface of the Earth is approximately 10 N/kg. Since the weight of an object is mg, increase in GPE is a special version of the equation $W = F \times d$, with $F = mg$ and $d = h$.

KINETIC ENERGY, KE

The kinetic energy of a moving object is calculated using the equation:

kinetic energy, KE (joules) = mass, m (kilograms) × speed squared, v^2 (metres squared per seconds squared)

$$KE = \frac{1}{2}mv^2$$

EXTENSION WORK

In Chapter 1 we saw the equation of uniformly accelerated motion:

$v^2 = u^2 + 2as$

This equation allows us to work out the final velocity, v, of an object that has a uniform acceleration, a, and that has accelerated through a distance, s, starting with an initial velocity of u.

If we drop an object from rest its initial velocity, $u = 0$, its acceleration is g m/s^2 and the distance that it falls is usually represented by h, so the equation becomes

$v^2 = 2gh$

Multiplying both sides of the equation by m, the mass of the object, gives:

$mv^2 = 2mgh$ which can be rearranged as $\frac{1}{2}mv^2 = mgh$

showing that the gain in KE of the object is equal to its loss of GPE.

We see that the amount of kinetic energy possessed by a moving object depends on its speed and its mass. As the Earth travels through space, orbiting the Sun, it runs the risk of colliding with chunks of matter that are drawn into the gravitational field of the Solar System. In fact, this is very common. If you have ever seen a shooting star – or, to give it its proper name, a meteor – you have seen the line of light produced as a small piece of space debris (waste) burns up on entering our atmosphere. This is an example of kinetic energy being transferred to heat and light by the friction produced between the air and the object passing through it.

▲ Figure 16.6 Meteors burn up on entering our atmosphere – we see them as 'shooting stars'.

EXTENSION WORK

The meteorite that caused the Arizona crater is thought to have hit the Earth travelling at 11 000 m/s and to have had a mass of 109 kilograms. It hit the ground with an energy equivalent to a 15 megaton hydrogen bomb, 1000 times greater than the atomic bomb dropped on Hiroshima at the end of the Second World War.

▲ Figure 16.7 This crater was created when a meteorite collided with Earth in Arizona.

EXAMPLE 3

Calculate the kinetic energy carried by a meteorite of mass 500 kg (less than that of an average-sized car) hitting the Earth at a speed of 1000 m/s.

$$KE = \tfrac{1}{2}mv^2$$

$$= \tfrac{1}{2} \times 500 \text{ kg} \times (1000 \text{ m/s})^2$$

$$= 250\,000\,000 \text{ J (or 250 MJ)}$$

CALCULATIONS USING WORK, GPE AND KE

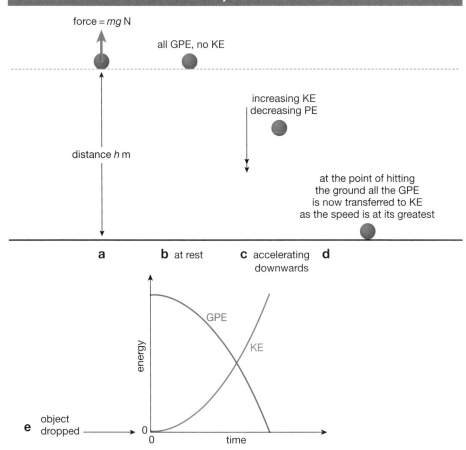

▲ Figure 16.8 GPE and KE of a falling object: **a** doing work to lift an object; **b** all GPE; **c** GPE transferring to KE during fall; **d** all KE at end of fall; **e** graph showing relationship between GPE and KE as the object falls

Work transfers energy to an object: $W = Fd$. An object of mass, m, weighs ($m \times g$) newtons so the force, F, needed to lift it is mg (Figure 16.8). If we raise the object through a distance h, the work done on the object is $mg \times h$. This is also the gain in GPE.

When the object is released, it falls – it loses GPE, but gains speed and so gains KE. At the end of the fall, all the initial GPE of the stationary (not moving) object has been transferred to the KE of the moving object. The graphs in Figure 16.8 show how the GPE of the object is changing into KE as it falls. The sum of the two graphs is always the same. Energy is conserved, so the loss of GPE is equal to the gain in KE.

work done lifting object = gain in GPE = gain in KE of the object just before hitting the ground

EXAMPLE 4

▲ Figure 16.9 A rollercoaster ride

In a rollercoaster ride the truck falls through a height of 17 m. Calculate the truck's speed at the bottom of this fall. (Take g = 10 N/kg)

If we assume that all the GPE of the truck at the top of the ride is transferred to KE at the bottom we can use the equation:

$$GPE = KE$$

$$mgh = \frac{1}{2}mv^2$$

The mass, m, of the truck appears on both sides of the equation, so it cancels out:

$$gh = \frac{1}{2}v^2$$

Substituting h = 17 m and g = 10 N/kg:

$$17 \text{ m} \times 10 \text{ N/kg} = \frac{1}{2}v^2$$

$$v^2 = 2 \times 17 \text{ m} \times 10 \text{ N/kg}$$

$$v = \sqrt{(2 \times 17 \text{ m} \times 10 \text{ N/kg})}$$

$$= 18.44 \text{ m/s (about 66 kph)}$$

Some of the GPE the truck had at the start will be transferred to heat and sound, so the speed at the bottom of the fall will be a little slower than this.

POWER

Power is the rate of transfer of energy or the rate of doing work.

James Watt (Figure 16.10a) is remembered as the inventor of the steam engine and is said to have been inspired by watching the lid on a kettle being forced up by the pressure of the steam forming inside. Neither story is accurate, but what is true is that Watt, working in partnership with Matthew Boulton, developed improvements to the steam engine that made it a commercial product and completely changed industry and transport.

▲ Figure 16.10 **a** James Watt (1736–1819) was a Scottish engineer who improved the performance of the steam engine and can be said to have started the Industrial Revolution – the beginning of the machine age. **b** A model of a steam engine that transfers the heat energy of steam to movement. Watt's engines were used to pump water out of mines.

The SI unit of power is named in honour of James Watt. The watt (W) is the rate of transfer or conversion of energy of one joule per second (1 J/s).

$$\text{power, } P \text{ (watts)} = \frac{\text{work done, } W \text{ (joules)}}{\text{time taken, } t \text{ (seconds)}}$$

$$P = \frac{W}{t}$$

ACTIVITY 1

▼ PRACTICAL: INVESTIGATE YOUR POWER OUTPUT

You may have done a simple experiment involving running upstairs to measure your output power. You do work as you raise your GPE, and to find your power output in watts you divide the work done by the time taken. The experiment is shown in Figure 16.11. Notice that calculating the work you do against gravity using force × distance works just as well as using the equation for *GPE* (mass × gravitational field strength × height).

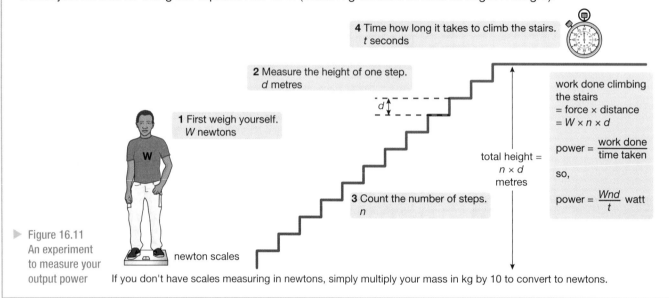

4 Time how long it takes to climb the stairs. *t* seconds

2 Measure the height of one step. *d* metres

1 First weigh yourself. *W* newtons

total height = *n* × *d* metres

3 Count the number of steps. *n*

work done climbing the stairs
= force × distance
= $W \times n \times d$

$$power = \frac{work\ done}{time\ taken}$$

so,

$$power = \frac{Wnd}{t}\ watt$$

Figure 16.11
An experiment to measure your output power

newton scales

If you don't have scales measuring in newtons, simply multiply your mass in kg by 10 to convert to newtons.

 Wear suitable footwear and only allow one person at a time on the staircase. Ensure the stairs are dry, in good condition and free of any obstacles.

A more convenient way of raising your GPE and getting to a higher floor in a building is to take a lift. The lift will transfer its energy input, usually electrical, to kinetic energy and then, if you are going up, to GPE. As usual, unwanted energy transfers are inevitable – sound and heat will be produced. If we know the weight of the lift and its contents and the height through which it moves, we can calculate the work done in the usual way. If we measure the time that the lift journey takes we can then calculate the power output of the lift motor. (Strictly this will be the useful power output – it will not take account of the wasted power due to unwanted energy transfers.)

EXAMPLE 5

If a lift and passengers have a combined weight of 4000 N and the lift moves upwards with an average speed of 3 m/s then what is the useful power output of the lift motor?

To keep the lift moving upwards at a steady speed, the lift motor must provide an upward force to balance the weight of the lift. This is 4000 N. In each second, this force is applied through a vertical distance of 3 m, so:

work done per second = 4000 N × 3 m/s

= 12 000 J/s

= 12 000 W

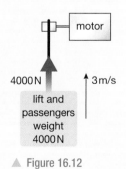

motor

4000 N   3 m/s

lift and passengers weight 4000 N

▲ Figure 16.12

CHAPTER QUESTIONS

More questions on work, power and efficiency can be found at the end of Unit 4 on page 168.

In the questions below, where necessary, take the strength of the Earth's gravity to be 10 N/kg.

SKILLS CRITICAL THINKING ⑤

1   James Joule showed that heat is a store of energy. He did this by showing that heat can be produced by using mechanical energy.

   a   Give an example of a process in which kinetic energy is transferred to heat.

   b   Describe how heat energy can be transferred to either kinetic energy or gravitational potential energy.

④

2   a   State the SI unit of work.

⑤

   b   Define the unit of work.

SKILLS PROBLEM SOLVING ⑦

   c   How much work is done in each of the following situations?
      i   A bag of six apples each weighing 1 N is lifted through 80 cm.
      ii   A rocket with a thrust of 100 kN travels to a height of 200 m.
      iii   A weightlifter raises a mass of 60 kg through a height of 2.8 m.
      iv   A lift of mass 200 kg lifts three people of mass 50 kg each through a distance of 45 m.

⑧

3   Water from a hydroelectric power station reservoir is taken from a reservoir (artificial lake) at a height of 800 m above sea level to turbines (engines) in the power station itself. The power station is at sea level. The reservoir holds 200 million (2×10^8) litres of water. If a litre of water has a mass of 1 kg, how much gravitational potential energy is stored in the water in the reservoir?

SKILLS CRITICAL THINKING

4   a   State how to calculate the kinetic energy stored by a moving object.

SKILLS PROBLEM SOLVING

   b   Work out the kinetic energy of the following:
      i   a man of mass 80 kg running at 9 m/s
      ii   an air rifle pellet of mass 0.2 g travelling at 50 m/s
      iii   a ball of mass 60 g travelling at 24 m/s.

⑨

5   A catapult fires a stone of mass 0.04 kg vertically upwards. If the stone has an initial kinetic energy of 48 J, how high will it travel before it starts to fall back to the ground?

6   If a coin is dropped from a height of 80 m, how fast will it be travelling when it hits the ground? State any assumptions you may need to make.

SKILLS CRITICAL THINKING ⑤

7   Define power and state its unit.

SKILLS PROBLEM SOLVING

8   A person with a mass of 40 kg runs upstairs in 12 s. The stairs have 20 steps and the height of each step is 20 cm.

⑥

   a   How much does the person weigh, in newtons?

⑤

   b   What is the total height that the person has climbed?

⑦

   c   Calculate how much work is done in climbing the stairs.

   d   What is the power output of the person running up the stairs?

9   A drag car, of mass 500 kg, accelerates from rest to a speed of 144 km/h in 5 s.

⑥

   a   What is its final speed in:
      i   m/h (metres per hour)
      ii   m/s?

⑧

   b   What is the increase in KE of the drag car?

   c   What is the average power developed by the drag car's engine?

PHYSICS ONLY

17 ENERGY RESOURCES AND ELECTRICITY GENERATION

In this chapter, you will learn about different resources of energy that are available to us on Earth. You will learn that some resources are renewable, while others cannot be replaced once used. As the demand for energy increases with the human population, there is a danger that non-renewable resources will run out. We therefore need to use fuel efficiently and to exploit more renewable resources.

▲ Figure 17.1 Humans consume vast amounts of fuel for transport, as well as heating and cooking.

LEARNING OBJECTIVES

■ Describe the energy transfers involved in generating electricity using:

■ wind
■ water
■ geothermal resources
■ solar heating systems

■ solar cells
■ fossil fuels
■ nuclear power

■ Describe the advantages and disadvantages of methods of large-scale electricity production from various renewable and non-renewable resources

The demand for energy increases all the time. The growth of the world population means more people need food and warmth. More people want to be able to travel. The fuels that we use to produce energy are being used up too quickly. We must use our remaining fuel supplies efficiently and look for new resources of energy. In Chapter 15, we saw how we can use energy more efficiently in heating our homes. We can make the energy supplies we have available last longer by being energy efficient.

We also need to understand what energy resources we have available on the Earth. In this chapter we shall look at different types of energy resource and the advantages and disadvantages of each resource for generating electricity. In particular we shall distinguish between renewable and non-renewable energy resources. New resources of energy are being researched all the time to meet our growing needs. We must also consider the effect that the use of different energy resources has on our environment. Some types of energy resource can cause long-term damage to our environment.

NON-RENEWABLE ENERGY RESOURCES

FOSSIL FUELS

▲ Figure 17.2 Fossil fuels include coal, oil and natural gas.

One of the main energy resources available on our planet is its supply of **fossil fuels**. Coal, oil and natural gas are all fossil fuels. They have been formed in the ground from dead vegetation or tiny creatures by a process that has taken millions of years. Once we have used them, it will take millions of years for new reserves of these fuels to be formed. Fossil fuels are, therefore, examples of non-renewable energy resources.

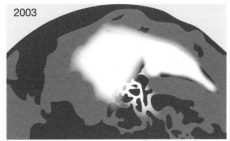

▲ Figure 17.3 Polar ice caps melting

A non-renewable energy resource is one that effectively cannot be replaced once it has been used.

Burning fossil fuels affects the environment, mainly by releasing carbon dioxide into the atmosphere. Carbon dioxide is a greenhouse gas. Greenhouse gases trap the Sun's heat in the Earth's atmosphere and cause the average temperature of the atmosphere to rise. This effect is called global warming and causes changes in the world's climate and melting of the polar ice caps. Burning coal releases more carbon dioxide into the atmosphere than burning oil or gas. Of the fossil fuels, natural gas produces the least carbon dioxide for the same energy output. There is no practical way of avoiding the release of carbon dioxide into the atmosphere when fossil fuels are burned, although some energy companies are researching ways of capturing it and storing it underground.

Most types of coal and oil contain some sulphur. When they are burned, this is converted to sulphur dioxide. Sulphur dioxide is then released into the atmosphere where it combines with water to form acid rain. Acid rain causes damage to people, plants and buildings. It is possible to remove the sulphur from these fuels but this increases the cost of the energy produced. It is also possible to remove sulphur dioxide from the waste gases when the fuel is burned, but this also increases the cost. International agreements are forcing

▲ Figure 17.4 William Perkin discovered the first synthetic dye in 1856 using substances produced from coal tar.

companies that emit large quantities of sulphur dioxide to clean up their waste gases, and acid rain is now less of a problem in Europe than it used to be. Acid rain is still a problem in many countries in the developing world.

Fossil fuels also provide valuable chemicals that can be used in the manufacture of a wide range of useful products (Figure 17.5). Once burned for energy production, these chemical resources are lost permanently. Burning such resources may be a very wasteful way of using them.

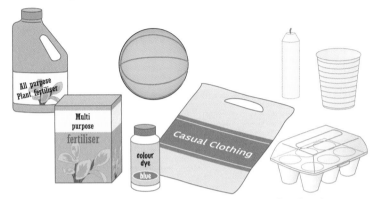

▲ Figure 17.5 Many products are manufactured using extracts from oil and coal.

NUCLEAR FUEL

Nuclear reactors use uranium to produce energy. For the nuclear process, a particular form or isotope (see page 223) of uranium is needed. Although a reactor only needs a small amount of uranium fuel, uranium is in limited supply. The uranium in the Earth was formed before the Solar System was formed, so once it has been used there will be no further supplies. It is, therefore, another example of a non-renewable resource.

Power generated from nuclear processes has the advantage of being 'clean'. It is clean because the process does not involve the production of greenhouse or other polluting gases. The cost per unit of electricity is very low, but nuclear power stations are expensive to build. The disadvantages of nuclear power are the risk of accidents and the problem of disposal of radioactive material once a power station is finished with it. Accidents that release radioactive materials like uranium and plutonium into the atmosphere pose long-lasting risks to living things.

EXTENSION WORK

'Fast breeder reactors' are so called because they create more nuclear fuel than they consume, but they still require uranium for their operation. The fuel produced is plutonium. This is extremely dangerous to life and is also a material used for the manufacture of nuclear weapons.

ELECTRICITY

Electricity is not an energy resource, because it has to be generated using other resources of energy. At present, most of the electricity used in the world is generated in power stations like the one shown in Figure 17.6.

Heat from nuclear fuel or from burning fossil fuels is used to heat water. This produces high-pressure steam that makes the blades of a turbine spin. A turbine is like a windmill or a fan, but with many more blades. The turbine is used to turn the generator, which generates the electricity (you will learn more about this in Chapter 21). The energy changes involved are shown in Figure 17.7.

Electricity can also be generated using renewable energy resources.

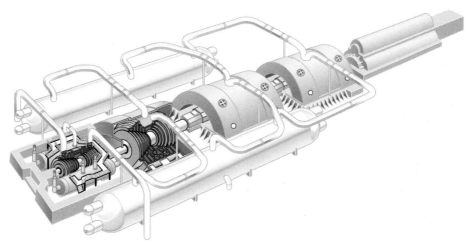

▲ Figure 17.6 Turbines and generators like these are used to produce electricity in power stations.

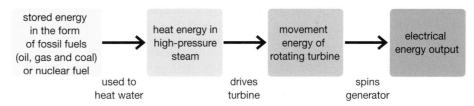

stored energy in the form of fossil fuels (oil, gas and coal) or nuclear fuel → heat energy in high-pressure steam → movement energy of rotating turbine → electrical energy output

used to heat water   drives turbine   spins generator

▲ Figure 17.7 Several energy changes are involved in producing electricity.

RENEWABLE ENERGY RESOURCES

A renewable energy resource is one that will not run out.

Wood is an example of a renewable energy resource. As wood is cut for fuel, new fast-growing trees are planted to replace those cut down. With careful management the supply of wood fuel can be maintained indefinitely. However, burning wood produces pollution and greenhouse gases. Wood is also more valuable if it is not burned, because it can be used in building or making furniture.

The demand for fuel and our worries about global warming and pollution have made us search for alternative energy resources. We need renewable energy resources that do not pollute the world or contribute to global warming.

HYDROELECTRIC POWER

▲ Figure 17.8 Moving water is a renewable energy resource.

The kinetic energy available in large quantities of moving water has been harnessed (used) for many hundreds of years. Water wheels have been used to transfer the energy stored by water in rivers to grind (break down) corn and power industrial machinery. A different kind of water wheel, called a turbine, is used to turn the generators in a **hydroelectric power** station. These power stations use the stored gravitational potential energy (GPE) of water in high reservoirs built in mountains. The GPE is transferred to kinetic energy (KE) as the water flows down the mountain to the power station below.

The energy transferred in this way is renewable. The Sun causes water to evaporate continuously and to be drawn up into the atmosphere. This water then falls as rain to be collected in reservoirs and used again. Moving water is a renewable resource.

Although hydroelectricity is a very clean, renewable resource, building reservoirs and power stations can spoil the landscape. The reservoir may also destroy or alter the natural habitat for wildlife.

TIDAL POWER

a

b

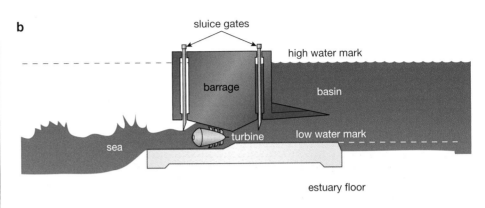

▲ Figure 17.9 **a** Tidal power station at La Rance, Brittany. **b** Tidal power also involves harnessing (using) the energy in moving water.

▲ Figure 17.10 The turbine has angled blades. As water is forced between the blades, they start to turn. The rotating turbine is connected to a generator to make it turn too.

The tides also involve the movement of huge amounts of water. Tidal power generation schemes, like that at La Rance in Brittany, generate power by turning turbines as the tide flows into a dammed (blocked) river estuary – where a river joins the sea. As the tide falls and the water flows out of the estuary the turbines turn again.

The energy for the movement of the tides is provided by the gravitational pull of the Moon and Sun. This is renewable energy using a small fraction of the continuous supply of gravitational energy.

There are not many places around the world suitable for building dams for tidal energy. If a dam is built, it affects the rise and fall of water in the estuary, and this is likely to damage habitats for wildlife. Some energy companies are now developing 'tidal stream turbines', which are like underwater wind turbines. These will be driven by tidal currents.

WAVE ENERGY

Energy can also be extracted from waves. The continuous movement of the surface of the seas and oceans is the result of a combination of tides and wind. A variety of methods have been developed to make use of the rise and fall of water due to waves. Figure 17.11 shows a system that has been developed to use the energy of waves. Again, this energy is renewable, as the movement energy of the waves is continuously available.

Water power is clean, producing no greenhouse gases or unwanted waste products.

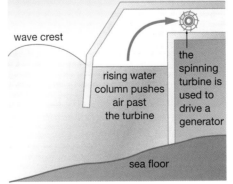
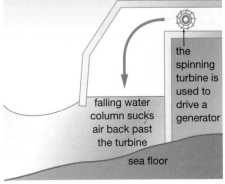

▲ Figure 17.11 An oscillating water column system for using wave energy

WIND POWER

▲ Figure 17.12 Old windmills were a way of using wind energy.

Winds are powered by the Sun's heat energy. Wind is a renewable source of energy that has been used for many centuries. Windmills have been used to grind corn and power machinery like pumps to drain (remove liquid from) lowland areas. Today, wind turbines drive generators to provide electrical energy.

The energy produced is clean, but wind power can only be harvested in regions where the wind blows with enough energy for a significant proportion of the year. Wind farms can cause environmental damage, as they change the appearance of the landscape. They also cause some noise pollution, and may kill birds and bats.

SOLAR POWER PRODUCING ELECTRICITY

Photovoltaic cells (solar cells) transfer light energy directly to electrical energy.

Photovoltaic (PV) cells are around 15% efficient. This means that 15% of the Sun's energy is transferred to useful electric energy. Improvements in PV efficiency and the fact that they are becoming cheaper to produce mean that more PV energy farms are being set up to provide large amounts of electrical energy. PV cells are also used to provide small amounts of electricity for use in places that cannot easily be connected to the electricity mains supply.

Figure 17.13 shows PV cells being used to power the Hubble space telescope, and a PV energy farm.

Advantage: They provide a renewable energy resource that does not produce greenhouse gases.

Disadvantage: Some people think that thousands of square metres of PV cells spoil the countryside. Others are worried that good farming land is lost when PV farms are built.

a    b

▲ Figure 17.13 Photovoltaic cells **a** on the Hubble space telescope, and **b** on an energy farm produce electricity when sunlight falls on them.

SOLAR POWER PRODUCING HEAT

▲ Figure 17.14 Solar heating panels used to heat water

Figure 17.14 shows panels that use heat from the Sun to warm water. They provide small quantities of hot water and reduce the amount of energy that might otherwise come from using non-renewable resources.

The design of efficient solar heating panels involves all the methods of heat transfer discussed in Chapter 15. Solar panels are often used in questions about heat transfer.

Solar heating panels absorb thermal radiation and use it to heat water. The panels are placed to receive the maximum amount of the Sun's energy. In the northern **hemisphere**, they must face south and be angled so that light falls on them as directly as possible for as long as possible. The structure of a typical solar heating panel is shown in Figure 17.15.

light

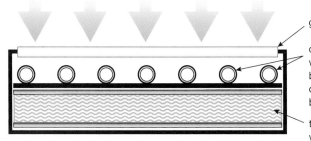

glass to trap an air layer

copper water pipes, with matt black surface, brazed to a sheet of copper with a matt black surface

thermal insulation layer faced with reflective foil on both upper and lower surfaces

▲ Figure 17.15 Here is one type of solar heating panel. They are designed to transfer as much energy to the water passing through them as possible.

Solar panels used to heat water (and therefore reduce energy used to heat water) are often used as an example of how a knowledge of conduction, convection and radiation can help to make these panels as efficient as possible. The use to generate electricity is described in the solar power boiler shown in Figure 17.16. Other methods are used to concentrate the heating effect of the Sun rays too.

Water is pumped through copper pipes brazed onto a copper sheet. Copper is used because it is an excellent thermal conductor. The surfaces of the sheet and the pipes have a matt (not shiny) black finish as this is the best absorber of heat radiation. The glass traps a layer of air above the copper to help insulate the unit and retain the heat. The backing is also designed to stop heat escaping to the surroundings. This kind of panel is reasonably efficient and the energy produced is more cost efficient than that from photovoltaic cells. Solar heating panels are used widely to provide water heating.

▲ Figure 17.16 A large number of mirrors focus heat from the Sun on a boiler in the tower at the centre. The steam produced drives a turbine to generate electricity.

Electricity can also be generated using solar heating. Curved mirrors are used to focus thermal radiation onto a boiler or pipes containing water to produce steam. The mirrors are controlled to reflect the Sun's heat onto the central tower throughout the day. The steam can be used to drive turbines which can be used to drive electricity generators.

GEOTHERMAL ENERGY

Geothermal energy is thermal energy stored deep inside the Earth. The heat in regions of volcanic activity was produced by the decay of radioactive elements like uranium. Volcanoes are evidence of the enormous heat and energy beneath the Earth's surface but do not provide a safe or reliable energy resource. However, heat from the ground can be used safely. In some areas of the world, like Iceland (Figure 17.17), geothermally heated water is readily available in springs and geysers. This is used to drive the turbines in electricity generation stations. The hot water is also used to provide domestic heating by sending it directly to houses. This resource is renewable, does not produce pollution and does not have a great impact on the environment. There are many areas of the world where geothermally heated springs can, and are, being used to provide energy.

▲ Figure 17.17 A geothermal electricity generating station in Iceland, with a huge swimming lagoon heated geothermally

SUPPLY AND DEMAND

Many different energy resources can be used to generate electricity, and they all have advantages and disadvantages. Some of the environmental disadvantages have already been discussed, but supply and demand also need to be taken into account. Different types of power station differ in the speed with which they can meet changes in the demand for electricity.

The demand for electricity varies from hour to hour, day to day and season to season. The way that the demand varies can be predicted to some extent. For example, there is a rise in demand in the early morning as people wake up, turn on lights and heaters and start to make breakfast, and of course in winter the demand for heat is much greater than in summer. Some sudden rises in demand are less predictable. A popular TV programme with an exciting episode can keep millions of viewers in front of their televisions – if they all decide to make a cup of tea as soon as the adverts come on, electricity usage will suddenly increase (electric kettles use a lot of power). The companies that supply electricity must be able to cope with these changes in demand, otherwise they are forced to cut off electricity to some consumers. This is not good for customer relations and, of course, means a reduction in the amount of electricity sold.

Nuclear power stations cannot be turned on instantly. The process of starting the fission reaction and heating up the core of the nuclear reactor is a long one. Clearly nuclear power stations cannot meet sudden variations in demand.

Power stations that burn fossil fuels can be started more quickly but can still take many hours to start producing electricity. Coal-fired stations take longer than oil-fired stations to develop the heat required to drive steam through the turbines. Gas-fired stations can respond most quickly to rises in demand.

Hydroelectric power stations provide a very reliable energy resource with the advantage of being able to respond very quickly to changes in the national demand for electricity. Unlike other types of power station, they are able to operate in reverse. This means that they can use extra electricity produced by other power stations that cannot be shut down quickly to pump water back up into the high-level reservoirs. This transfers the electrical energy back to gravitational potential energy, which can then be re-transferred when needed at a later time. This is the only realistic way of 'storing' large amounts of extra electrical energy.

The US Department of Energy has proposed the building of a nuclear waste storage facility beneath the Yucca Mountains. High-level nuclear waste will be placed in concrete containers that are supposed to hold the nuclear radiation safely for 300 to 1000 years.

Wind power is dependent on the strength, direction and frequency of wind. Although wind farms are located in windy areas, they cannot be relied upon to produce electricity at the times when it is needed most. Tidal power is not available continuously, but the times at which it will be available are predictable.

COST

Planners must also look at the financial costs of electricity generation. Nuclear power uses a relatively cheap fuel. Uranium produces huge amounts of energy, so the cost of energy per unit of fuel used is low. However, building a nuclear power station is very expensive. Nuclear power requires complex technology and very high standards of safety. On top of these 'start-up' costs, planners must also consider the expense of decommissioning (closing down) a nuclear power station at the end of its useful working life. For conventional power stations, this is a routine demolition job, but for nuclear power stations the task is not as straightforward. Radioactive materials must be handled with great care and stored in a way to ensure that none escape.

We see that, although the running costs of nuclear power are relatively low, the pay-back time is very long. (The pay-back time is how long it takes for the income from selling electricity to cover the cost of building the power station.)

The cost of setting up a wind farm is much lower than the cost of building a nuclear power station and there are no fuel costs. However, the amount of energy produced by wind farms is comparatively low. The pay-back time for wind generators is therefore quite long.

LOOKING AHEAD

Sometimes physics is described as the study of matter and energy. As we have stated in this unit, energy, work and power affect us in everyday life. We use energy to move things around, keep us warm and to manufacture things. As the population of the world increases the demand for energy also increases. Fossil fuels are a non-renewable resource and, although new reserves may be discovered and new methods of extracting these reserves found, they will run out eventually.

Energy is needed to move an object against a force. You will learn more about the forces that govern the way the Universe works: gravity, electromagnetic and the two types of nuclear force. By understanding how to break the forces that hold the atom together, physicists have made power from nuclear fission. Understanding how to force atoms close enough together for them to form more massive atoms will provide us with nuclear power from fusion reactors.

Building experimental fusion reactors is a huge challenge for engineers and scientists. If successful they may provide a long-term solution to the world's demand for energy without the problems of nuclear waste – but a fusion reactor capable of delivering energy on the scale we need has not yet been built.

Scientists continue to search for new solutions to producing energy and to improving the efficiency of the devices we use to cut down on energy waste.

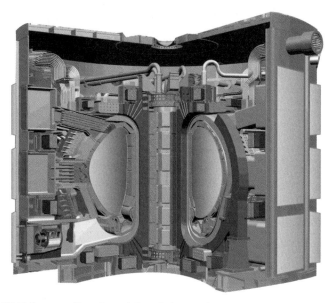

▲ Figure 17.18 Cross-section of a prototype fusion reactor

CHAPTER QUESTIONS

More questions on energy resources can be found at the end of Unit 4 on page 168.

SKILLS   CRITICAL THINKING

1 Here is a list of ways of ways of producing electricity:

    A wind farms   E gas-fired power stations

    B coal-fired power stations   F tidal power stations

    C hydroelectric power stations   G using geothermal energy

    D nuclear power stations

 a State which of these produces 'clean' electricity.

 b State which of these uses 'fossil' fuels.

      c State which of these uses a renewable energy resource.

      d State which of these produces waste that remains dangerous to people and animals for very long periods of time.

SKILLS   REASONING

2 From the list in Question 1 choose those methods of producing electricity that are only suitable in certain places and state the type of location required for each method.

SKILLS   CRITICAL THINKING

3 Nuclear power generation does not produce greenhouse gases. This is a significant advantage of nuclear power over power stations that burn fossil fuels.

      a Describe two other advantages of using nuclear power.

      b Describe two disadvantages of using nuclear power.

4 The demand for electricity varies on an hourly and a daily basis. Sometimes the changes in demand are predictable, but sometimes they are not. Give two examples of predictable changes in demand and two examples of changes in demand that are not predictable.

5 Electricity is generated by a number of different types of power station. Give three examples of different types of power station and compare how well they are able to respond to sudden changes in demand.

END OF PHYSICS ONLY

UNIT QUESTIONS

a A fluorescent lamp is 25% efficient. Choose the statement that correctly describes how the lamp performs.

    A the lamp only works for a quarter of the time

    B 25% of the electrical energy transferred to the lamp is wasted

    C the lamp does not work properly

    D 25% of the energy supplied to the lamp is transferred to light   **(1)**

b A car uses energy stored in chemical form in petrol. When the engine is running this energy is transferred to other types of stored energy. Which of the following energy stores is useful?

    A noise

    B heat

    C movement

    D hot exhaust gases   **(1)**

c Which of the following materials is a good thermal conductor?

    A glass

    B aluminium

    C air

    D water   **(1)**

d Which of the following statements about thermal radiation is false?

    A It cannot travel through glass.

    B It can travel through a vacuum.

    C It travels at the speed of light.

    D It is absorbed by black objects better than shiny ones.   **(1)**

e Four students, A, B, C and D, measure their power output by running upstairs and timing how long it takes. Which of the four students has the greatest power output?

    A mass of 50 kg, takes 20 s to gain a height of 10 m

    B mass of 40 kg, takes 30 s to gain a height of 15 m

    C mass of 45 kg, takes 10 s to gain a height of 5 m

    D mass of 55 kg, takes 25 s to gain a height of 15 m   **(1)**

(Total for Question 1 = 5 marks)

2   An electric motor is used to raise a load that weighs 800 N through a distance of 30 m.

a How much work does it do in raising the load?   **(3)**

b Calculate the power output of the motor if it takes 16 s to raise the load.   **(2)**

c The motor is 75% efficient.

   i Explain what this means.   **(2)**

   ii Calculate the electrical power that must be supplied to the motor to raise this load in the time stated. Give your answer in kW.   **(2)**

d Draw a labelled Sankey diagram to represent the energy transfers that take place as the motor raises the load.   **(3)**

(Total for Question 2 = 12 marks)

3   The diagram shows a toy rocket launcher that uses a spring in a tube.

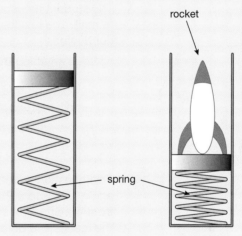

rocket

spring

a Here are statements that describe the events leading to the launch of the toy rocket:

1 spring released

2 rocket leaves launcher

3 rocket gains kinetic energy

4 spring stores elastic potential energy

5 child does work squashing the spring

6 spring transfers stored energy to rocket

Choose from the following list the letter that describes the events in the order that they must happen.   **(1)**

A 1 5 6 3 2 4   **B** 5 4 1 6 3 2   **C** 1 5 4 6 3 2   **D** 5 4 1 6 2 3

b Describe the energy transfers that take place after the rocket leaves the launcher, up to and including when it hits the ground.   **(4)**

(Total for Question 3 = 5 marks)

4 A student wants to show that water is a poor conductor of heat. The diagram below shows the investigation he sets up.

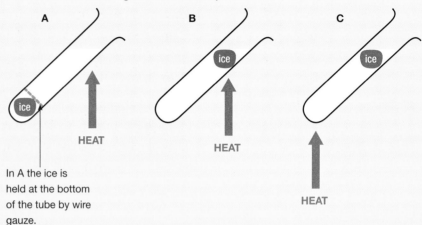

In A the ice is held at the bottom of the tube by wire gauze.

The student measures how long it takes for each piece of ice in the water to melt. The results will help him to show whether or not water is a poor conductor.

a State anything the student should do to make sure this is a fair test. **(3)**

b i Assuming that this is a fair test, in which order do you think the ice will melt, starting with the slowest? Give an explanation for your answer. **(4)**

ii How can this investigation show that water is a poor conductor? **(3)**

(Total for Question 4 = 10 marks)

5 This question is about PV (photovoltaic) cells. A student wanted find the best angle to set up a PV cell to get the maximum amount of energy transferred from the Sun's rays to electrical energy. Here is the apparatus she set up and the results she obtained.

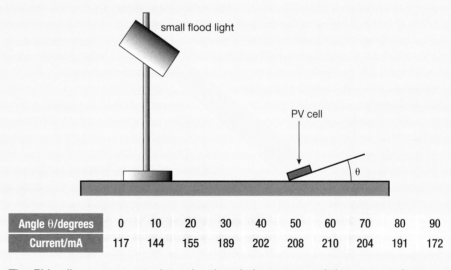

Angle θ/degrees	0	10	20	30	40	50	60	70	80	90
Current/mA	117	144	155	189	202	208	210	204	191	172

The PV cell was connected to a load and she measured the current, *I*, produced by the cell with the cell at different angles, θ, to the horizontal.

a State:

i the dependent variable **(1)**

ii the independent variable **(1)**

iii a control variable. **(1)**

b On a graph grid, 18 cm by 22 cm, plot a graph of the current produced by the PV cell against the angle of the cell to the horizontal.   **(5)**

c Use the graph to find:

   i any anomalous result   **(1)**

   ii the angle to get the maximum output from the PV cell in the set-up shown.   **(1)**

d When installing PV panels it is important to have them pointing towards the Sun as you would expect. In the experimental set-up shown, the lamp is modelling conditions in the UK in early March at midday (when the Sun reaches its highest position in the sky).

   i How could the experiment be set up to model different positions of Sun?   **(1)**

   ii The experiment is repeated by another student at a time and place when the Sun is directly overhead at midday. Sketch a line on your graph to show how this student's results might appear. Label this line 'Sun overhead'.   **(3)**

(Total for Question 5 = 14 marks)

UNIT 5
SOLIDS, LIQUIDS AND GASES

Matter can exist in three basic forms: as a solid, a liquid or a gas. In this photo all three are present, though we can only see the solid iceberg and the liquid ocean. Fortunately the iceberg is surrounded by gas, the mixture of oxygen and nitrogen that makes up our atmosphere. Our atmosphere usually contains some water in gas form, which we cannot see. When gaseous water in the atmosphere condenses into tiny water droplets we can see these as clouds.

18 DENSITY AND PRESSURE

One way of characterising materials is by their density. This chapter looks at density and at how materials can affect things around them by exerting pressure.

▲ Figure 18.1 Some effects of density and pressure

LEARNING OBJECTIVES

■ Know and use the relationship between density, mass and volume:

$$\text{density} = \frac{\text{mass}}{\text{volume}} \quad \rho = \frac{m}{V}$$

■ Practical: investigate density using direct measurements of mass and volume

■ Know and use the relationship between pressure, force and area:

$$\text{pressure} = \frac{\text{force}}{\text{area}} \quad p = \frac{F}{A}$$

■ Understand how the pressure at a point in a gas or liquid at rest acts equally in all directions

■ Know and use the relationship for pressure difference:

pressure difference = height × density × gravitational field strength

$$p = h \times \rho \times g$$

UNITS

In this section you need to use degree Celsius (°C) and Kelvin (K) as the units of temperature (these units both represent the same change in temperature but the Kelvin scale starts from absolute zero, as explained later), kilogram per cubic metre (kg/m^3) as the unit of density, cubic metre (m^3) as the unit of volume, pascal (Pa) as the unit of pressure and square metre (m^2) as the unit of area. It is important to remember $1\ m^3 = 1\,000\,000\ cm^3$ (a cubic metre is 1 million cubic centimetres) and $1\ m^2 = 10\,000\ cm^2$ (a square metre is 10 thousand square centimetres).

KEY POINT

ρ is the Greek letter rho (pronounced 'row') and is the usual symbol for density. This is not to be confused with p which represents pressure.

The properties of a material affect how it behaves, and how it affects other materials around it. The balloon can fly because the gas inside it has a very low density. The skis spread the weight of the skier over the snow so she does not sink into it. The submersible is designed to explore the seabed – it has a very strong hull to withstand the high pressure from water deep in the oceans.

DENSITY

Solids, liquids and gases have different properties and characteristics. One such characteristic is density. Solids are often very dense – that is, they have a high mass for a certain volume. Liquids are often less dense than solids, and gases have very low densities.

The density (ρ) of a material can be calculated if you know the mass (m) of a certain volume (V) of the material, using this equation:

$$\text{density}, \rho = \frac{\text{mass}, m}{\text{volume}, V}$$

$$\rho = \frac{m}{V}$$

The units for density depend on the units used for mass and volume. If mass is measured in kilograms and volume in cubic metres, the units for density are kilograms per cubic metre (kg/m³). Density can also be measured in grams per cubic centimetre (g/cm³).

A piece of iron has a mass of 390 kg and a volume of 0.05 m³. What is its density?

$$\text{density} = \frac{\text{mass}}{\text{volume}}$$

$$= \frac{390 \text{ kg}}{0.05 \text{ m}^3}$$

$$= 7800 \text{ kg/m}^3$$

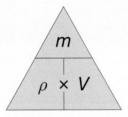

▲ Figure 18.2 The equation for density can be rearranged using the triangle method.

HINT

If an examination question asks you to write out the equation for calculating density, mass or volume, always give the actual equation in words or using standard symbols, such as $\rho = \frac{m}{V}$.

ACTIVITY

▼ **PRACTICAL: INVESTIGATE THE DENSITY OF SOLIDS**

The density of a substance can be determined by measuring the mass and volume of a sample of the material, and then calculating the density.

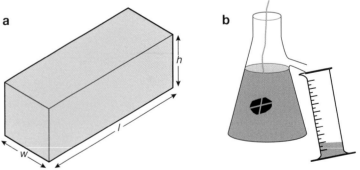

▲ Figure 18.3 **a** The volume of a regular solid can be calculated by multiplying its length (*l*), width (*w*) and height (*h*). **b** The volume of an irregular object can be determined using a displacement can and a measuring cylinder.

Use a half-metre rule to measure the length, width and height of a regular solid. When measuring the volume of the liquid that is displaced (pushed out) by an irregular solid make sure that the measuring cylinder is on a level (horizontal) surface and that you look at the scale straight on as shown (to avoid **parallax error**).

You then need to measure the mass of the solid. Use weighing scales for this.

EXAMPLE 2

The mass of 50 cm³ of a liquid and a measuring cylinder is 146 g. The mass of the empty measuring cylinder is 100 g. What is the density of the liquid in kg/m³?

mass of 50 cm³ of liquid = 146 g – 100 g

= 46 g

= 0.046 kg

50 cm³ = 0.000 05 m³

$$\rho = \frac{m}{V}$$

$$= \frac{0.046 \text{ kg}}{0.000\,05 \text{ m}^3}$$

$$= 920 \text{ kg/m}^3$$

or

$$= \frac{46 \text{ g}}{50 \text{ cm}^3}$$

$$= 0.92 \text{ g/cm}^3$$

$$= 920 \text{ kg/m}^3$$

HINT

Check the units when you are working out density – don't mix up g and m³, for example. Also check which units the question asks for in your final answer.

1 kg = 1000 g

To convert kg to g, divide by 1000.

1 m³ = 1 000 000 cm³

To convert cm³ to m³, divide by 1 000 000 (or 10⁶).

Alternatively, work out the density in g/cm³, then multiply your answer by 1000.

PRESSURE UNDER A SOLID

You can push a drawing pin into a piece of wood quite easily, but you cannot make a hole in the wood with your thumb, no matter how hard you push! The small point of the drawing pin concentrates all your pushing force into a tiny area, so the pin goes into the wood easily. Similarly, it is easier to cut things with a sharp knife than a blunt one, because with a sharp knife all the force is concentrated into a much smaller area.

Pressure is defined as the force per unit area. Force is measured in newtons (N) and area is measured in square metres (m²). The units for pressure are pascals (Pa), where 1 Pa is equivalent to 1 N/m².

Pressure (p), force (F) and area (A) are linked by the following equation:

$$\text{pressure, } p \text{ (pascals)} = \frac{\text{force, } F \text{ (newtons)}}{\text{area, } A \text{ (square metres)}}$$

$$p = \frac{F}{A}$$

REMINDER

When answering questions, start with the equation you need to use in words or recognised symbols.

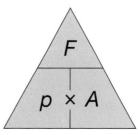

▲ Figure 18.4 The equation for pressure can be rearranged using the triangle method.

EXAMPLE 3

An elephant has a weight of 40 000 N, and her feet cover a total area of 0.1 m².

A woman weighs 600 N and the total area of her shoes in contact with the ground is 0.0015 m². Who exerts the greatest pressure on the ground?

Elephant:

$$p = \frac{F}{A}$$

$$= \frac{40\,000\ \text{N}}{0.1\ \text{m}^2}$$

$$= 400\,000\ \text{Pa (or 400 kPa)}$$

Woman:

$$p = \frac{F}{A}$$

$$= \frac{600\ \text{N}}{0.0015\ \text{m}^2}$$

$$= 400\,000\ \text{Pa (or 400 kPa)}$$

They both exert an equal pressure.

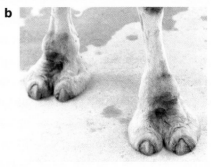

▲ Figure 18.5 **a** The caterpillar tracks on this vehicle spread its weight over a large area.
b Camels have large feet so they are less likely to sink into loose sand.

Some machines, including cutting tools like scissors, bolt cutters and knives, need to exert a high pressure to work well. In other applications, a low pressure is important. Tractors and other vehicles designed to move over mud have large tyres that spread the vehicle's weight. The pressure under the tyres is relatively low, so the vehicle is less likely to sink into the mud. Caterpillar tracks used on bulldozers and other earth-moving equipment serve a similar purpose. (In Figure 18.5a the caterpillar tracks are very large.)

PRESSURE IN LIQUIDS AND GASES

The submersible shown in Figure 18.1 has a very strong hull to withstand the high pressure **exerted** on it by seawater. Pressure in liquids acts equally in all directions, as long as the liquid is not moving. You can easily demonstrate this using a can with holes punched around the bottom, as shown in Figure 18.6. When the can is filled with water, the water is forced out equally in all directions.

Gases also exert pressure on things around them. The pressure exerted by the atmosphere on your body is about 100 000 Pa (although the pressure varies slightly from day to day). However, the pressure inside our bodies is similar, so we do not notice the pressure of the air.

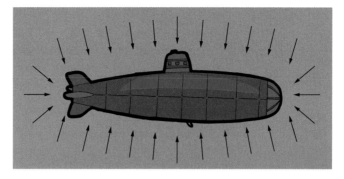 

▲ Figure 18.6 Pressure in liquids acts equally in all directions. A can of water with holes can be used to demonstrate this.

One of the first demonstrations of the effects of air pressure was carried out by Otto van Guericke in 1654, in Magdeburg, Germany. Van Guericke had two large metal bowls made, put them together and then pumped the air out. The bowls could not be pulled apart, even when he attached two teams of horses to the bowls.

▲ Figure 18.7 Van Guericke's experiment at Magdeburg

You can do the same experiment in the laboratory, using much smaller bowls called Magdeburg hemispheres. When air is inside the spheres, the pressure is the same inside and outside. If the air is sucked out, pressure is only acting from the outside. The hemispheres cannot be pulled apart until air is let back into them.

a

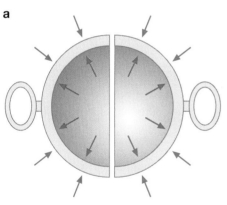

b

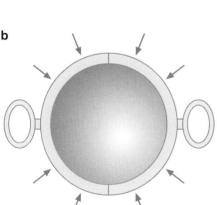

▲ Figure 18.8 **a** When the hemispheres are full of air, the forces are the same inside and outside. **b** When the air is taken out, there is only a force on the outside of the hemispheres.

EXAMPLE 4

A laboratory set of Magdeburg hemispheres has a surface area of 0.045 m². What is the total force on the outside of the hemispheres?

$$F = p \times A$$
$$= 100\,000 \text{ Pa} \times 0.045 \text{ m}^2$$
$$= 4500 \text{ N}$$

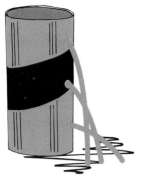

▲ Figure 18.9 Pressure in a liquid increases with depth.

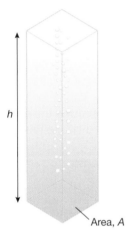

▲ Figure 18.10 We can work out the pressure difference between two points in a liquid by considering a column of the liquid.

PRESSURE AND DEPTH

The experiment shown in Figure 18.9 demonstrates that the pressure in a liquid increases with depth. We can work out the pressure difference by thinking about a column of water, as shown in Figure 18.10.

The force at the bottom of the column is equal to all the weight of water above it. The volume of this water (V) is found by multiplying the area of its base (A) by the height (h) of the column. We can work out the mass (m) of the water by multiplying the volume ($A \times h$) by the density (ρ).

$$\text{mass of water, } m = (A \times h) \times \rho$$

The force (F) on the bottom of the water column is equal to the weight of this volume of water, which is the mass, m ($A \times h \times \rho$), multiplied by the gravitational field strength (g) (see page 34).

$$F = (A \times h \times \rho) \times g$$

As we are concerned with the pressure on the base of the column, we divide the force by the area:

$$F = \frac{A \times h \times \rho \times g}{A}$$

The area of the column therefore does not matter, and we can calculate the pressure difference between two points in a liquid using the equation:

pressure difference, p (Pa) = height, h (m) × density, ρ (kg/m³) × gravitational field strength, g (N/kg)

$$p = h \times \rho \times g$$

Be careful not to muddle the symbol for pressure, p, with the symbol for density, ρ.

This equation can be used for calculating pressure differences in other liquids or gases, as long as you know their densities.

EXAMPLE 5

The first experiments on air pressure were carried out using a simple barometer. This was made by filling a long glass tube with mercury, and then inverting it into a bowl of mercury. The mercury falls until the weight of the column is supported by the pressure of the air on the mercury in the bowl.

If the column of mercury is 0.74 m high, what is the air pressure?

The density of mercury is 13 600 kg/m³. Take the gravitational field strength as 10 N/kg.

$$p = h \times \rho \times g$$

$$= 0.74 \text{ m} \times 13\,600 \text{ kg/m}^3 \times 10 \text{ N/kg}$$

$$= 100\,640 \text{ Pa}$$

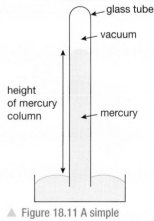

▲ Figure 18.11 A simple mercury barometer

The unit for pressure is named after Blaise Pascal (1623–1662). He was the first person to demonstrate that air pressure decreases with height. He persuaded his brother-in-law to take a mercury barometer up a mountain and measure the air pressure at different stages in the climb. He also arranged for someone at the base of the mountain to measure the air pressure at different times during the day. The measurements showed that the pressure of the air does decrease as you climb.

LOOKING AHEAD – WEATHER AND PRESSURE

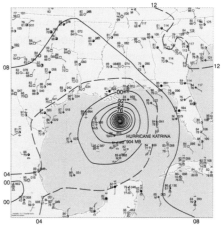

▲ Figure 18.12 Weather maps show isobars (lines of equal air pressure). The centre of hurricane Katrina is a very low pressure region surrounded by isobars that are very close together. The result of the pressure difference is very strong winds.

▲ Figure 18.13 Tornados regularly cause huge amounts of damage in many parts of the world.

Variations in atmospheric pressure are partly responsible for the weather systems around the world. Meteorologists produce maps showing how atmospheric pressure varies from place to place to help them make predictions about future weather systems. Regions of low pressure produce winds as air flows from surrounding areas of high pressure.

Sometimes the wind speeds cause violent weather systems with tornadoes and hurricanes causing extensive damage and loss of life. Physics provides important tools to enable meteorologists to make more accurate predictions about when and where these devastating events will occur.

CHAPTER QUESTIONS

More questions on density and pressure can be found at the end of Unit 5 on page 193.

SKILLS   CREATIVITY, REASONING

1 A Greek scientist called Archimedes was asked to check the purity of the gold in a crown. He did this by comparing the density of the crown with the density of pure gold.

a Describe how he could have measured the density of an irregular-shaped object such as a crown.

SKILLS ▶ PROBLEM SOLVING

b Pure gold has a density of 19 000 kg/m³. Suppose the crown had a volume of 0.0001 m³. What mass should the crown be if it is made of pure gold?

2 People working on the roofs of buildings often lay a ladder or plank of wood on the roof. They walk on the ladder rather than the roof itself.

SKILLS ▶ REASONING

a Explain why using a ladder or plank will help to prevent damage to the roof.

SKILLS ▶ PROBLEM SOLVING

b A workman's weight is 850 N, and each of his boots has an area of 210 cm². Calculate the maximum pressure under his feet when he is walking. Give your answer in pascals.

c The workman lays a plank on the roof. The plank has an area of 0.3 m² and a weight of 70 N. What is the maximum pressure under the plank when he is walking on it?

HINT

Remember that all your weight is on one foot at some point while you are walking. To convert cm² to m², divide by 10 000 or 10⁴.

3 A manometer can be used to find the pressure of a gas. The difference in the level of the liquid in each side of the tube indicates the difference in pressure on the water surface at A and the pressure of the gas on the water surface at B. (Assume atmospheric pressure is 100 kPa.)

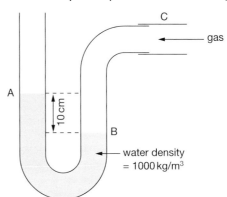

a What is the difference in pressure in Pa between A and B? Use $g = 10$ N/kg.

b What is the pressure of the gas in tube C?

SKILLS ▶ REASONING

c Describe what will happen when the gas is turned off and the pipe is removed from C.

19 SOLIDS, LIQUIDS AND GASES

All matter is made up of particles that are continuously moving. The arrangement and movement of the particles determine the properties of the material. In gases, scientists have discovered laws that describe the relationship between pressure, temperature and volume. In this chapter, you will learn what these laws are and how this relationship can be explained in terms of the behaviour of the particles.

▶ Figure 19.1 The botanist Robert Brown (1773–1858) made an observation while studying pollen grains under a microscope that has become known as Brownian motion in his honour. The observation led to the kinetic theory of matter that we use to explain the behaviour of solids, liquids and gases.

LEARNING OBJECTIVES

PHYSICS ONLY

- Explain why heating a system will change the energy stored within the system and raise its temperature or produce changes of state

- Describe the changes that occur when a solid melts to form a liquid, and when a liquid evaporates or boils to form a gas

- Describe the arrangement and motion of particles in solids, liquids and gases

- Practical: obtain a temperature–time graph to show the constant temperature during a change of state

- Know that specific heat capacity is the energy required to change the temperature of an object by one degree Celsius per kilogram of mass (J/kg °C)

- Use the equation:

  change in thermal energy = mass × specific heat capacity × change in temperature

  $$\Delta Q = m \times c \times \Delta T$$

- Practical: investigate the specific heat capacity of materials including water and some solids

- Explain how molecules in a gas have random motion and that they exert a force and hence a pressure on the walls of a container

- Understand why there is an absolute zero of temperature which is −273 °C

- Describe the Kelvin scale of temperature and be able to convert between the Kelvin and Celsius scales

- Understand why an increase in temperature results in an increase in the average speed of gas molecules

- Know that the Kelvin temperature of a gas is proportional to the average kinetic energy of its molecules

- Explain, for a fixed amount of gas, the qualitative relationship between:

  - pressure and volume at constant temperature

  - pressure and Kelvin temperature at constant volume

- Use the relationship between the pressure and Kelvin temperature of a fixed mass of gas at constant volume:

  $$\frac{p_1}{T_1} = \frac{p_2}{T_2}$$

- Use the relationship between the pressure and volume of a fixed mass of gas at constant temperature:

  $$p_1 V_1 = p_2 V_2$$

PHYSICS ONLY

We think that all matter is made up of tiny particles that are moving. The way that the particles are arranged and the way that they move determine the properties of a material, such as its state at room temperature or its density.

THE STATES OF MATTER

Some of the properties of a substance depend on the chemicals it is made from. However, substances can exist in different states. The main states of matter are solid, liquid and gas. We are used to finding some substances in each state in everyday life – for example, water is familiar as ice, as water and as steam. There are other substances that we usually see in only one state – for example, we rarely see iron in any state other than solid, or experience oxygen in any state other than as gas.

Substances can change state by the processes of melting, evaporation, boiling, freezing and condensing.

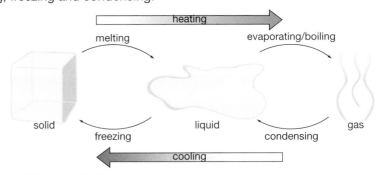

▲ Figure 19.2 Changes of state

PROPERTIES OF THE DIFFERENT STATES OF MATTER

Most of this chapter will be about the properties of gases, but we shall mention solids and liquids for completeness.

SOLIDS

Solids have a definite rigid shape and they are often very dense. The density of a substance is a measure of how tightly packed the particles are.

Some solids have high densities because the particles that they are made from are very closely packed in a regular arrangement. There are strong forces between the particles, which give solid objects their definite shape and, in some materials, a great deal of strength.

Although the particles are held together by strong forces, they can still move. They vibrate about their fixed positions in the solid. When we supply energy to a solid by heating it, the particles vibrate more – they move more quickly. We notice the increase in the kinetic energy of the particles in a substance as an increase in the temperature of the substance.

LIQUIDS

Liquids share a property with gases – they have no definite shape. However, the particles that make up liquids tend to stick together, unlike gas particles. Liquids will occupy the lowest part of any container but gases will expand to fill any container that they are in. Liquids have much greater densities than gases. This is because the particles in liquids are still very close together, like they are in solids. Because the particles in liquids are close together, they still attract one another and hold together. In liquids, there is no fixed pattern and the particles can move around more freely than in solids. As we heat liquids, the movement of the particles becomes more energetic.

GASES

In gases the particles are very spread out, with large spaces between them. This means that the forces holding them together are small. Gases have very low densities and no definite shape. Gases can also be squashed into a smaller space (compressed). Particles of a gas are moving randomly all the time. The particles will bump into anything in the gas, or into the walls of the container, and the forces caused by these collisions are responsible for the pressure that gases exert.

Solids and liquids are very difficult to compress because the particles in them are almost as close together as they can be.

SUMMARY OF THE PROPERTIES OF SOLIDS, LIQUIDS AND GASES

Property	Solids	Liquids	Gases
definite shape	yes	no	no
can be easily compressed	no	no	yes
relative density	high	high	low
can flow (fluid)	no	yes	yes
expands to fill all available space	no	no	yes

a
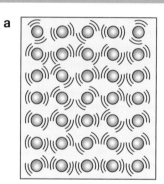
solid

b
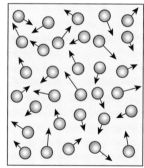
liquid

c
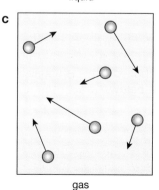
gas

▲ Figure 19.3 The different arrangements of the particles in **a** solids; **b** liquids and **c** gases (not drawn to scale)

The particles (molecules) in a solid (Figure 19.3a):

- are tightly packed
- are held in a fixed pattern or crystal structure by strong forces between them
- vibrate around their fixed positions in the structure.

The particles (molecules) in a liquid (Figure 19.3b):

- are tightly packed (still very close together, like in solids)
- are not held in fixed positions but are still bound together by strong forces between them
- move at random with no fixed positions.

The particles (molecules) in a gas (Figure 19.3c):

- are very spread out
- have no fixed positions and the forces between them are very weak
- move with a rapid, random motion.

MEASURING HEAT ENERGY

We have seen that when heat energy is supplied to solids, liquids and gases it increases the kinetic energy of their molecules, which we detect as an increase in temperature. Different substances need different amounts of heat energy to cause the same increase in temperature.

We define the specific heat capacity (s.h.c.), c, of a substance as the amount of energy required to increase the temperature of 1 kilogram of that substance by 1 °C. The unit of s.h.c. is J/kg °C.

Δ, the Greek letter delta, means a change in the quantity that follows it. θ, the Greek letter theta, is used to represent temperature so $\Delta\theta$ (delta theta) means a change in temperature. If we cool something down then $\Delta\theta$ is negative and so, therefore, is ΔQ. This simply means that, to cool something down, we need to remove thermal energy from it.

We use the following equation to work out how much energy is needed to change the temperature of an object by a given amount:

change in thermal energy, ΔQ (joules) = mass, m (kilograms) × specific heat capacity, c × change in temperature, $\Delta\theta$ (°C)

$$\Delta Q = m \times c \times \Delta\theta$$

EXAMPLE 1

If you fill a kettle with 300 g of water at an initial temperature of 15 °C, how much energy is needed to make the water heat up to boiling? The s.h.c. of water is 4200 J/kg °C.

Write down what you know:

m = 0.3 kg (you must work in consistent units)

c = 4200 J/kg °C

and, as the boiling point of water is 100 °C, $\Delta\theta$ = 85 °C

Write down the equation you are using (no marks for this as it is a given equation):

$$\Delta Q = m \times c \times \Delta\theta$$

Substitute the correct values into the equation (this will normally attract a method mark).

$$\Delta Q = 0.3 \text{ kg} \times 4200 \text{ J/kg °C} \times 85 \text{ °C}$$

Now complete the calculation and include the correct unit in your answer.

$$\Delta Q = 107\,100 \text{ J}$$

This figure assumes that all the thermal energy provided goes to raise the temperature of the water. In practice some energy will be used to heat up the kettle and some will be lost to the surroundings through thermal radiation, thermal conduction and convection and some water will evaporate before the boiling point is reached.

ACTIVITY 1

▼ **PRACTICAL: INVESTIGATE THE SPECIFIC HEAT CAPACITY OF A SUBSTANCE**

Rearranging the equation above to make c the subject we get:

$$c = \frac{\Delta Q}{m \times \Delta\theta}$$

So we must:

- measure the mass, m, in kg of the substance under test using electronic scales

- measure the initial temperature and the final temperature, using a thermometer to find $\Delta\theta$ °C

- determine the amount of thermal energy supplied – this is usually done with an electric immersion heater as shown below.

! The immersion heater (and, eventually, the metal block) will get hot enough to burn the skin. If used in water, the immersion heater must not have its top immersed.

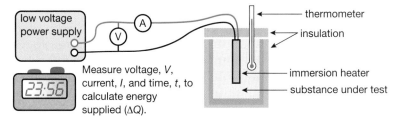

Measure voltage, V, current, I, and time, t, to calculate energy supplied (ΔQ).

▲ Figure 19.4 Apparatus for measuring the specific heat capacity of a substance

Figure 19.4 shows typical apparatus for finding the s.h.c. of a metal. A cylinder of the metal is drilled to allow an electric heater and a thermometer to be inserted as shown. Use a balance to find the mass of the cylinder and a thermometer to measure its temperature. Turn the heater on and measure the voltage, V (volts), supplied to the heater, the current drawn, I (amps), and the length of time, t (seconds), the heater is on. Make a note of the highest temperature reached and calculate $\Delta\theta$, the rise in temperature.

The thermal energy supplied by the heater is $\Delta Q = V \times I \times t$ (page 65).

Substitute the measured values for m, ΔQ and $\Delta\theta$ into the equation shown above and calculate the s.h.c.

The insulation is used to cut down heat losses to the surroundings.

This method can be used to measure the s.h.c. of water, with a weighed amount (m) of water in beaker (with a lid and insulation). The water can be heated with an immersion heater and heat energy given to the water calculated in the same way ($\Delta Q = VIt$). The temperature increase is measured with a thermometer.

EXAMPLE 2

In an experiment to measure the specific heat capacity of aluminium, a cylinder of mass 0.5 kg, initially at 20 °C, was supplied with 16 500 J of thermal energy. At the end of this heating the temperature had risen to 50 °C. Calculate the s.h.c. of aluminium.

Write down what you know:

$m = 0.5$ kg

$\Delta Q = 16\,500$ J

temperature rise from 20 °C to 50 °C is $\Delta\theta = 30$ °C

Substitute these values into the given equation:

$$c = \frac{\Delta Q}{m \times \Delta\theta}$$

$$= \frac{16\,500 \text{ J}}{0.5 \text{ kg} \times 30 \text{ °C}}$$

and calculate the value:

$c = 1100$ J/kg °C

As heat will still be lost and some of the heat energy supplied heats the immersion heater and the thermometer, then this answer is larger than the expected value of 900 J/kg °C

EXAMPLE 3

To determine the specific heat capacity of water, 1 litre of water (mass, m, = 1 kg) at a starting temperature of 20 °C is poured into an electric kettle with a power of 2.4 kW. The water in the kettle just reaches boiling point after 2½ minutes. Use this information to calculate the s.h.c. of water.

Do you think your answer will be too big or too small? Give a reason for your answer.

As power is the rate of supply of energy you can calculate the energy transferred to the water (ΔQ) using:

energy transferred, ΔQ (J) = power, p (W) × time, t (seconds) (see page 65)

$$\Delta Q = 2400\ W \times 150\ s$$
$$= 360\,000\ J$$

The temperature increase is 80 °C and, conveniently, the mass of water, m, is 1 kg.

Substitute these values into the given equation:

$$c = \frac{\Delta Q}{m \times \Delta \theta}$$

$$= \frac{360\,000\ J}{1\ kg \times 80\ °C}$$

and calculate the value:

$$c = 4500\ J/kg\,°C$$

The answer will be higher than the accepted value because of heat losses to the surroundings and the energy that is used to heat the kettle itself.

THE ENERGY INVOLVED IN CHANGE OF STATE

When you supply energy to a substance you would expect its temperature to rise and this is generally true.

ACTIVITY 2

▼ PRACTICAL: INVESTIGATE TEMPERATURE DURING A CHANGE OF STATE

The following two experiments show situations where this is not the case.

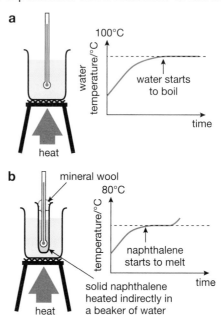

▲ Figure 19.5 **a** Heating water; **b** Heating solid naphthalene

Figure 19.5 shows two experiments involving a change of state. In Figure 19.5a water is being heated continuously. The water temperature rises until the water starts to boil and then remains constant until all the water has been turned from liquid to gas. The energy supplied is doing work separating the water molecules. In Figure 19.5b naphthalene (used to make moth balls), which has a melting point of 80 °C, is being heated continuously in a beaker of water. The temperature of the solid naphthalene rises until it starts to melt. Until all the naphthalene has melted, its temperature remains constant. The temperature will start to rise again when all the naphthalene is in the liquid state.

The thermal energy supplied is used to overcome the forces between the molecules of the liquid in the case of water and between the molecules of the solid in the case of naphthalene. If we allow the naphthalene to cool down, its temperature will drop until the naphthalene starts to turn back into a solid and will not start to fall again until all the liquid has turned back to solid. During any change of state, melting or freezing (solidifying), boiling or condensing (turning back from gas to liquid), the temperature remains constant.

END OF PHYSICS ONLY

THE GAS LAWS

We are now going to focus our attention on the properties of gases. We shall explain the different properties in terms of the movement of particles.

We have already said that gases are made up of molecules that are moving. We believe that the particles in gases are spread out and constantly moving in a random, haphazard way.

When the molecules hit the walls of a container they exert a force. The combined effect of the huge number of collisions results in the pressure that is exerted on the walls of the container.

BOYLE'S LAW

▲ Figure 19.6 Robert Boyle (1627–1691) was an Anglo-Irish chemist and philosopher. As well as discovering the law that bears his name, he worked with Robert Hooke at Oxford developing the air pump.

The scientist Robert Boyle (Figure 19.6) discovered something that you have probably noticed if you have ever used a bicycle pump: air is squashy! He noticed that you can squeeze air in a cylinder and that it springs back to its original volume when you release it. You can try this for yourself with a plastic syringe (Figure 19.7).

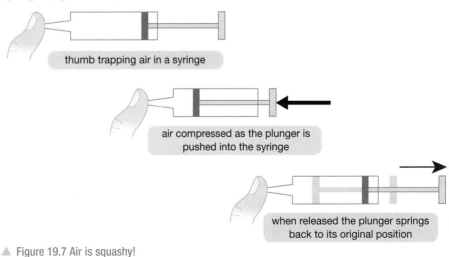

thumb trapping air in a syringe

air compressed as the plunger is pushed into the syringe

when released the plunger springs back to its original position

▲ Figure 19.7 Air is squashy!

▲ Figure 19.8 A modern version of Boyle's experiment to see how the volume of a gas depends on the pressure exerted on it.

> ⚠ Eye protection is needed in case any tubing leaks or detaches, which will cause gas under high pressure to be ejected.

Boyle devised an experiment to see how the volume occupied by a gas depends on the pressure exerted on it. Pressure is the force acting per unit area. This is measured in N/m². One N/m² is called a pascal (Pa). A version of Boyle's experiment is shown in Figure 19.8.

Boyle took care to make sure that the trapped gas he was studying stayed at the same temperature. He increased the pressure on the gas and made a note of the new volume. His results looked like the graph shown in Figure 19.9a.

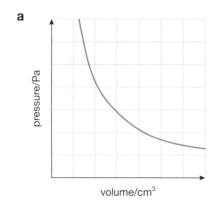
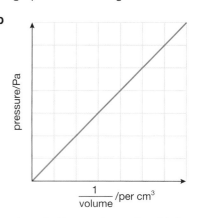

▲ Figure 19.9 **a** Graph to show how the pressure of a gas at constant temperature varies with the volume; **b** Graph of pressure against $\dfrac{1}{\text{volume}}$

Boyle noticed that when he doubled the pressure, the volume of the gas halved. If we plot pressure (p), against $\dfrac{1}{\text{volume}}$ $\left(\dfrac{1}{V}\right)$, as in Figure 19.9b, we can see from the straight line passing through the origin that p is proportional to $\dfrac{1}{V}$.

This discovery, called Boyle's law, is expressed in the equation:

$$p_1 V_1 = p_2 V_2$$

This means that if you take a fixed mass of gas that has a pressure p_1 and a volume V_1 and change either the pressure or the volume, the new pressure p_2 multiplied by the new volume V_2 will be the same as $p_1 \times V_1$, if the temperature of the gas remains the same.

EXAMPLE 4

Atmospheric pressure is 100 kPa (100 000 Pa). Some air in a sealed container has a volume of 2 m³ at atmospheric pressure. What would be the pressure of the air if you reduced its volume to 0.2 m³?

First write down what we know:

$$p_1 = 100 \text{ kPa}$$
$$V_1 = 2 \text{ m}^3$$
$$V_2 = 0.2 \text{ m}^3$$
$$p_1 V_1 = p_2 V_2$$
$$p_2 \times 0.2 \text{ m}^3 = 100 \text{ kPa} \times 2 \text{ m}^3$$
$$p_2 = \frac{100 \text{ kPa} \times 2 \text{ m}^3}{0.2 \text{ m}^3}$$
$$= 1000 \text{ kPa}$$

HINT

You can do all your working in kPa if you wish, rather than converting the pressure to pascals. You can also use cubic centimetres for volume if these units are given in the question. It does not matter which units you use, as long as you use the same ones throughout the question.

Gases can be compressed because the gas molecules are very spread out. When a gas is squashed into a smaller container it presses on the walls of the container with a greater pressure. This is explained in terms of particle theory as follows.

If the gas is kept at the same temperature, the average speed of the particles stays the same. (Remember that temperature is an indication of the kinetic energy of the particles.) If the same number of particles is squeezed into a smaller volume, they will hit the container walls more often. Each particle exerts a tiny force on the wall with which it collides. More collisions per second means a greater average force on the wall and, therefore, a greater pressure.

ABSOLUTE ZERO

Boyle took care to conduct his experiment at constant temperature. He was aware that temperature also had an effect on the pressure of a gas. Figure 19.10 shows an experiment to investigate how the pressure of a gas depends on its temperature.

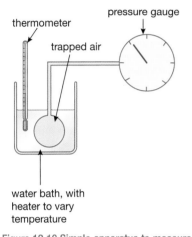

thermometer

pressure gauge

trapped air

water bath, with heater to vary temperature

▲ Figure 19.10 Simple apparatus to measure the pressure of a fixed volume of gas at a range of temperatures

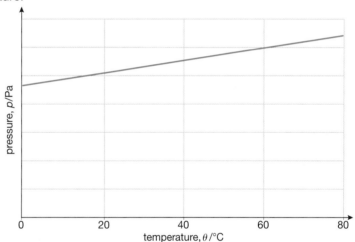

▲ Figure 19.11 Graph to show how the pressure of a fixed volume of gas varies with temperature

The gas is kept at constant volume, and its pressure is measured at a range of temperatures. The graph in Figure 19.11 shows typical results for this experiment.

The graph shows that the pressure of the gas increases as the temperature increases. We could also say that the gas pressure gets smaller as the gas is cooled.

What happens if we keep cooling the gas? The graph in Figure 19.12 shows this.

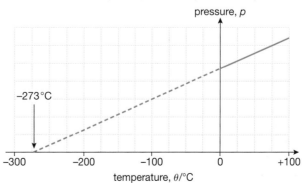

▲ Figure 19.12 At a temperature of −273 °C, the pressure of the gas would be zero. This temperature is known as 'absolute zero'.

As we cool the gas, the pressure keeps decreasing. The pressure of the gas cannot become less than zero. This suggests that there is a temperature below which it is not possible to cool the gas further. This temperature is called absolute zero. Experiments show that absolute zero is approximately −273 °C.

The Kelvin temperature scale starts from absolute zero. The Kelvin temperature of a gas is proportional to the average kinetic energy of its molecules.

To convert a temperature on the Celsius scale (in °C) to a Kelvin scale temperature (in K), add 273 to the Celsius scale temperature:

- temperature in K = temperature in °C + 273
- temperature in °C = temperature in K − 273

EXAMPLE 5

a At what temperature does water freeze, in Kelvin?

Water freezes at 0 °C. To convert 0 °C to Kelvin:

$$T = (0\,°C + 273)\,K$$
$$= 273\,K$$

b What is room temperature, in Kelvin?

Typical room temperature is 20 °C, so:

$$T = (20\,°C + 273)\,K$$
$$= 293\,K$$

c What temperature is 400 K on the Celsius scale?

$$\theta = (400\,K - 273)\,°C$$
$$= 127\,°C$$

If we redraw the graph of pressure against temperature using the absolute or Kelvin temperature scale, we get a graph that is a straight line passing through the origin, as shown in Figure 19.13. This shows that the pressure of the gas is proportional to its Kelvin temperature. For example, if you heat a gas from 200 K (−73 °C) to 400 K (127 °C), its pressure will double.

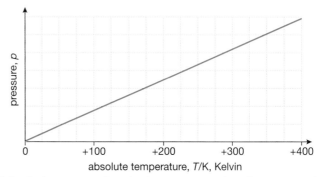

▲ Figure 19.13 Graph of pressure against absolute temperature for a fixed amount of gas at constant volume

We often use the symbol θ for temperature in °C. For temperature in K we always use the symbol T. Remember to convert Celsius temperatures to Kelvin when using gas law equations in which temperature change occurs.

For a fixed mass of gas at a constant volume:

$$\frac{p_1}{T_1} = \frac{p_2}{T_2}$$

NB T must be measured in Kelvin.

EXAMPLE 6

You take an empty tin and put the lid on tightly. You heat it using a Bunsen burner until the temperature of the air inside is 50 °C. What is the pressure of the air inside the tin? The temperature of the room is 20 °C.

First write down what you know:

p_1 = 100 kPa (this is atmospheric pressure)

The starting temperature is 20 °C, but we must change this to Kelvin, so:

T_1 = 20 + 273 K = 293 K.

The end temperature, 50 °C, must also be converted so:

T_2 = 50 + 273 K = 323 K

$$\frac{p_1}{T_1} = \frac{p_2}{T_2}$$

$$p_2 = p_1 \times \frac{T_2}{T_1}$$

$$= 100 \text{ kPa} \times \frac{323 \text{ K}}{293 \text{ K}}$$

$$= 110 \text{ kPa}$$

EXTENSION WORK

Cooling things further becomes more and more difficult as they get closer to 0 K. Scientists have managed to cool gases down to within a few thousandths of a degree above absolute zero. At this temperature, even gases like helium and hydrogen can be liquefied. The gas laws do not apply once a gas is either cooled or compressed (forced into a small volume) to the point that they become liquid.

The relationship can be explained as follows.

The number of gas particles and the space, or volume, they occupy remain constant. When we heat the gas the particles continue to move randomly, but with a higher average speed. This means that their collisions with the walls of the container are harder and happen more often. This results in the average pressure exerted by the particles increasing.

When we cool a gas the kinetic energy of its particles decreases. The lower the temperature of a gas the less kinetic energy its particles have – they move more slowly. At absolute zero the particles have no thermal or movement energy, so they cannot exert a pressure.

CHAPTER QUESTIONS

More questions on solids, liquids and gases can be found at the end of Unit 5 on page 193.

PHYSICS ONLY

SKILLS ▸ INTERPRETATION

SKILLS ▸ CRITICAL THINKING

1 a Draw up a table to summarise the way that particles are arranged in solids, liquids and gases.

  b How do these different arrangements of particles explain the physical properties of solids, liquids and gases?

2 Copy and complete the following paragraph about the particle theory of matter.

Matter is made up of _____ that are in continuous _____. When we supply heat energy to matter the _____ move _____. Gases exert a pressure on their containers because the _____ are continually _____ with the walls. The pressure will_____ when the gas is heated in a container of fixed volume because the _____ are moving _____.

END OF PHYSICS ONLY

 **SKILLS** CRITICAL THINKING

3 Explain how ideas about particles can account for the absolute zero of temperature.

SKILLS PROBLEM SOLVING

4 a Convert the following Celsius temperatures to Kelvin temperatures.
  i 0 °C
  ii 100 °C
  iii 20 °C

  b Convert the following Kelvin temperatures to Celsius temperatures.
  i 250 K
  ii 269 K
  iii 305 K

SKILLS ANALYSIS

5 State what happens in the situations shown in the diagram. Explain your answers using ideas about particles.

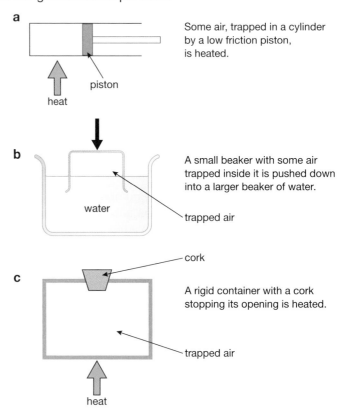

a Some air, trapped in a cylinder by a low friction piston, is heated.

piston

heat

b A small beaker with some air trapped inside it is pushed down into a larger beaker of water.

water

trapped air

cork

c A rigid container with a cork stopping its opening is heated.

trapped air

heat

PHYSICS ONLY

SKILLS PROBLEM SOLVING

6 a A fixed amount of gas occupies a volume of 500 m³ when under a pressure of 100 kPa. Calculate the volume that this amount of gas will occupy when its pressure is increased to 125 kPa. You should show each stage of your calculation.

SKILLS REASONING

  b What do you need to assume about the temperature of the gas? Explain your answer.

 **SKILLS** PROBLEM SOLVING

7 a You put the lid on a jar in a kitchen where the air temperature is 20 °C. The jar gets left out in the winter, and the air temperature drops to –5 °C. Explain what happens to the air pressure in the cold jar.

SKILLS REASONING

  b Explain why it might be more difficult to remove the lid of the jar when it is cold.

END OF PHYSICS ONLY

UNIT QUESTIONS

SKILLS   PROBLEM SOLVING   **1**

a   A rectangular block of metal of density 8000 kg/m³ measures 2 cm by 3 cm by 10 cm. What is the mass of the block?

   A 48 g

   B 0.48 kg

   C 4.8 kg

   D 48 kg   **(1)**

b   Which of the following will exert the greatest pressure when standing on the ground?

   A an elephant: weight 50 000 N, area of one foot 0.2 m²

   B atmospheric pressure: 10⁵ Pa

   C ballerina: weight 450 N, on one point, area 0.0015 m²

   D man: weight 800 N, area of one foot 0.025 m²   **(1)**

SKILLS   CRITICAL THINKING   **5**

c   Which of the following descriptions most accurately describes a gas?

   A very low density, easily compressed, cannot flow

   B high density, cannot be compressed, definite shape

   C very low density, easily compressed, expands to fill available space

   D high density, difficult to compress, can flow   **(1)**

d   The specific heat capacity of a liquid is

   A the amount of heat energy required to make 1 kilogram of the liquid boil

   B the amount of heat energy required to raise the temperature of 1 kilogram of the liquid by 1 °F

   C the amount of heat energy required to raise the temperature of 1 kilogram of the liquid by 1 °C

   D the amount of heat energy required to raise the temperature of 1000 cm³ of the liquid by 1 °C   **(1)**

e   Boyle's law states that

   A if you double the pressure on a gas its volume doubles

   B if you double the pressure on a gas its temperature halves

   C if you double the pressure on a gas its volume doubles provided its temperature is constant

   D if you double the pressure on a gas its volume halves provided its temperature is constant   **(1)**

(Total for Question 1 = 5 marks)

SKILLS   INTERPRETATION   **2**   Look at the diagram below showing a small gas bubble in a glass of fizzy drink, like cola.

a   Copy the diagram and use eight arrows to shows how the pressure of the liquid acts on the gas bubble.   **(2)**

b Describe what happens to the bubble as it floats up towards the top of the drink and explain the reason for your answer (assume that the amount of gas in the bubble does not change as the bubble rises up through the cola). **(3)**

(Total for Question 2 = 5 marks)

3 The diagram shows part of the apparatus used by a student to find the density of a stone.

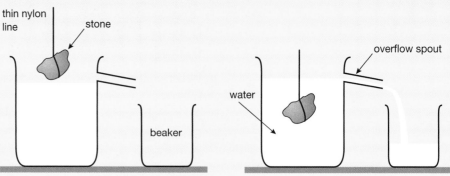

a **i** What is the student measuring? **(1)**

ii What other piece of apparatus is required to complete the measurement? **(1)**

iii What precautions should the student take to ensure that the measurement is accurate? **(1)**

b **i** State the equation that the student needs to calculate the density of the stone. **(1)**

ii What other measurement must be made? State the name of the apparatus needed to make this measurement. **(1)**

(Total for Question 3 = 5 marks)

4 The diagram shows a simple mercury barometer made by filling a strong glass tube with mercury and turning it upside down (with the open end of the tube closed) in a dish of mercury and then opening the end below the level of mercury in the dish. Mercury runs out of the tube until the pressure of the mercury at X is equal to the pressure of the air acting on the surface of the mercury in the dish.

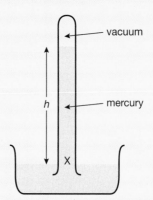

The height, h, of mercury in the tube is a measure of air (atmospheric) pressure. A typical value for h is 76 cm.

The density of mercury is 13 600 kg/m³. Use this value to calculate the atmospheric pressure in Pa (pascals). **(3)**

(Total for Question 4 = 3 marks)

PHYSICS ONLY

SKILLS > INTERPRETATION

5 A beaker filled with water initially at 20 °C is cooled in a freezer until its temperature drops to –20 °C (the temperature inside the freezer). The temperature in the beaker is measured with a digital thermometer and recorded at regular intervals until it reaches –20 °C.

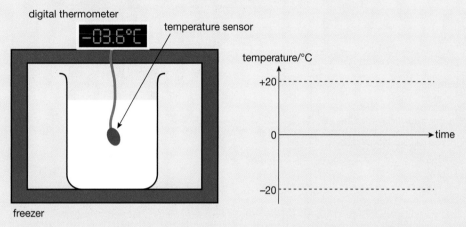

a Copy the graph axes shown and sketch how the temperature of the contents of the beaker changes with time from its starting temperature of +20 °C until it reaches –20 °C. **(5)**

SKILLS > ANALYSIS

b Referring to your sketch graph explain what happens to the water as it is cooled from 20 °C to –20 °C. **(5)**

(Total for Question 5 = 10 marks)

SKILLS > PROBLEM SOLVING

6 An electric kettle rated at 2.5 kW of mass 0.5 kg contains 1.5 kg of water at 20 °C. The kettle is turned on and the kettle and its contents are heated to 100 °C.

The s.h.c. (specific heat capacity) of the metal the kettle is made of is 500 J/kg °C and the s.h.c. of water is 4200 J/kg °C.

a Calculate the total amount of heat energy needed. **(5)**

b How long must the kettle be switched on for to heat the kettle and water to 100 °C? Give your answer in minutes to two significant figures. **(4)**

SKILLS > REASONING

c In practice it will take longer than this to boil the water in the kettle. Explain why. **(1)**

HINT

You may need to refer to Chapter 2 to find the relationship between electrical power, energy and time.

(Total for Question 6 = 10 marks)

END OF PHYSICS ONLY

UNIT 6
MAGNETISM AND ELECTROMAGNETISM

In Norse (Norwegian) mythology the spectacular light display we call the aurora borealis was seen as a bridge of fire that connected the Earth to the sky. Nowadays we know it is caused by charged particles, emitted by the Sun, interacting with the Earth's magnetic field and atmosphere. In this section we will be looking at magnetism and the crucial role it plays in our everyday lives.

20 MAGNETISM AND ELECTROMAGNETISM

There are two types of magnets that we use in our everyday lives. These are permanent magnets and electromagnets. A permanent magnet has a magnetic field around it all the time. The strength and direction of this field is not easy to change. An electromagnet when turned on also has a magnetic field around it but its strength and direction can be changed very easily. In this chapter you will learn about the factors affecting the magnetic field around an electromagnet and how electromagnets are used in several important devices.

▲ Figure 20.1 Electromagnets can be used to lift iron or steel objects.

The huge electromagnet in Figure 20.1 is being used in a scrapyard to pick up large objects that contain iron or steel. When the objects have been moved to their new position the electromagnet is turned off and the objects fall.

LEARNING OBJECTIVES

- Know that magnets repel and attract other magnets and attract magnetic substances

- Describe the properties of magnetically hard and soft materials

- Practical: investigate the magnetic field pattern for a permanent bar magnet and between two bar magnets

- Understand the term magnetic field line

- Know that magnetism is induced in some materials when they are placed in a magnetic field

- Describe how to use two permanent magnets to produce a uniform magnetic field pattern

- Know that an electric current in a conductor produces a magnetic field around it

PHYSICS ONLY

- Describe the construction of electromagnets

- Draw magnetic field patterns for a straight wire, a flat circular coil and a solenoid when each is carrying a current

UNITS

In this section you will need to use ampere (A) as the unit of current, volt (V) as the unit of voltage and watt (W) as the unit of power.

MAGNETISM AND MAGNETIC MATERIALS

Magnets are able to attract objects made from magnetic materials such as iron, steel, nickel and cobalt. Magnets cannot attract objects made from materials such as plastic, wood, paper or rubber. These are non-magnetic materials.

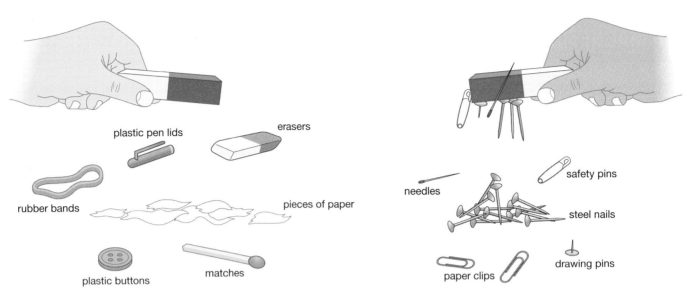

plastic pen lids

erasers

rubber bands

pieces of paper

plastic buttons

matches

needles

safety pins

steel nails

paper clips

drawing pins

▲ Figure 20.2 Magnets attract some objects and not others.

MAGNETS

The strongest parts of a magnet are called its poles. Most magnets have two poles. These are called the north pole and the south pole.

If two similar poles are placed near to each other they repel. If two dissimilar (opposite) poles are placed near to each other they attract.

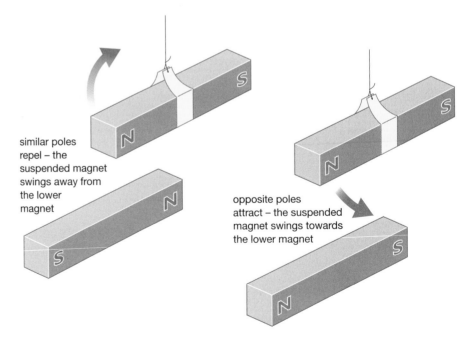

similar poles repel – the suspended magnet swings away from the lower magnet

opposite poles attract – the suspended magnet swings towards the lower magnet

▲ Figure 20.3 Similar poles repel and opposite poles attract.

Permanent magnets like the bar magnets shown in Figure 20.3 are made from a magnetically hard material such as steel. A magnetically hard material keeps its magnetism once it has been magnetised. Iron is a magnetically soft material and is not suitable for a permanent magnet. Magnetically soft materials lose their magnetism easily and are therefore useful as temporary magnets.

MAGNETIC FIELDS

Around every magnet there is a volume of space where we can detect magnetism. This volume of space is called a **magnetic field**. Normally a magnetic field cannot be seen but we can use iron filings or plotting compasses to show its shape and discover something about its strength and direction.

ACTIVITY 1

▼ PRACTICAL: INVESTIGATE THE MAGNETIC FIELD PATTERNS OF BAR MAGNETS

1 Place a bar magnet between two books and place a sheet of paper or thin card over it.

2 Sprinkle some iron filings on the paper above the magnet.

3 Tap the paper very gently.

4 The iron filings will move to show the magnetic field pattern.

OR

1 Place a bar magnet on a piece of paper.

2 Place a large number of small compasses on the paper near the magnet.

3 Look carefully at the pattern shown by the needles of the compasses.

Repeat the same experiment using two bar magnets, placing them about 5 cm apart.

> **!** Avoid skin contact with iron filings, since they irritate the skin. Do not blow them off the paper, because they may get into the eyes, which is very painful.

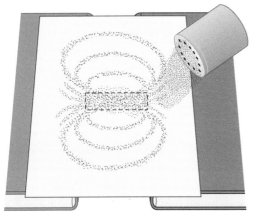

if you make a pencil dot at the end of the needle in each position of the compass, you can plot the field line

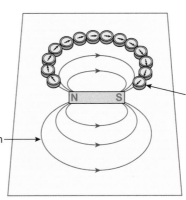

in each position, the needle of the plotting compass lines up with the field line, pointing from north to south

▲ Figure 20.4 We can see the shape of the magnetic field around a magnet by using iron filings or plotting compasses.

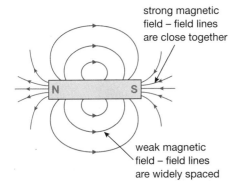

strong magnetic field – field lines are close together

weak magnetic field – field lines are widely spaced

▲ Figure 20.5 The magnetic field around a bar magnet follows a pattern like this.

We draw magnetic fields like that in Figure 20.5 using magnetic field lines. Magnetic field lines don't really exist but they help us to visualise the main features of a magnetic field.

The magnetic field lines:

▪ show the shape of the magnetic field

▪ show the direction of the magnetic force – the field lines 'travel' from north to south

▪ show the strength of the magnetic field – the field lines are closest together where the magnetic field is strongest.

OVERLAPPING MAGNETIC FIELDS

If two magnets are placed near each other, their magnetic fields overlap and affect each other. We can investigate this using iron filings or plotting compasses. Figure 20.6 shows the different field patterns we would see.

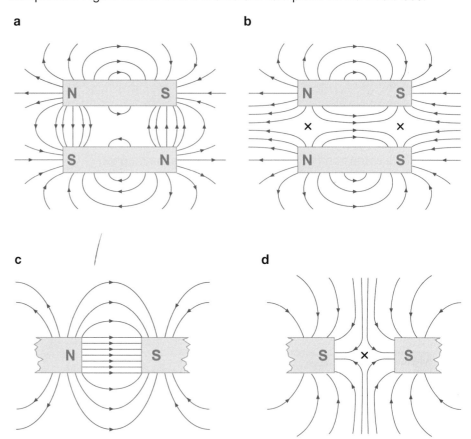

▲ Figure 20.6 Magnetic fields around pairs of magnets

CREATING A UNIFORM MAGNETIC FIELD

If we look very carefully at the magnetic field created between the north pole of one magnet and the south pole of a different magnet we will see that the field is shown as a series of straight field lines that are evenly spaced. A field like this is described as a uniform magnetic field – that is, its strength and direction is the same everywhere. Creating a uniform field like this is extremely useful as we will see later in this chapter and in the next chapter.

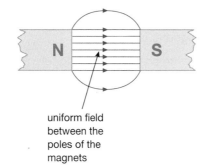

uniform field between the poles of the magnets

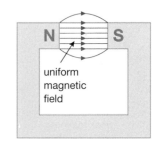

uniform magnetic field

u-shaped permanent magnets can be used to create a uniform magnetic field

▲ Figure 20.7 Two examples of uniform magnetic fields

INDUCED MAGNETISM

If we place an object made from a magnetic material, for example, an iron nail, inside a magnetic field it becomes a magnet. We say that magnetism has been induced in the nail. Because iron is a magnetically soft material, its induced magnetism is temporary and disappears if the permanent magnet is removed. If the nail is made from a magnetically hard material such as steel it will retain some of its magnetism after the magnet is removed.

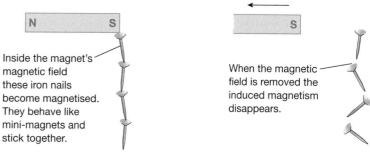

Inside the magnet's magnetic field these iron nails become magnetised. They behave like mini-magnets and stick together.

When the magnetic field is removed the induced magnetism disappears.

▲ Figure 20.8 Introducing magnetism in iron nails

PHYSICS ONLY

ELECTROMAGNETISM

When there is a current in a wire a magnetic field is created around it. This is called electromagnetism. The field around the wire is quite weak and circular in shape.

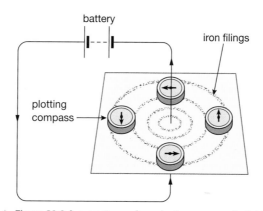

The shape and direction of the magnetic field around a current-carrying wire can be seen using iron filings and plotting compasses. Changing the direction of the current, changes the direction of the magnetic field.

▲ Figure 20.9 A current-carrying wire has a magnetic field around it.

EXTENSION WORK

The direction of the magnetic field depends upon the direction of the current. It can be found using the right-hand grip rule (for fields).

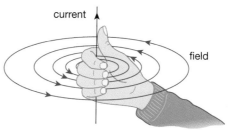

With the thumb of your right hand pointing in the direction of the current, your fingers will curl in the direction of the field.

▲ Figure 20.10 You can work out the direction of the field using the right-hand grip rule (for fields).

If the wire is made into a flat, single-turn coil (circular wire), the magnetic field around the wire changes shape and is as shown in Figure 20.11.

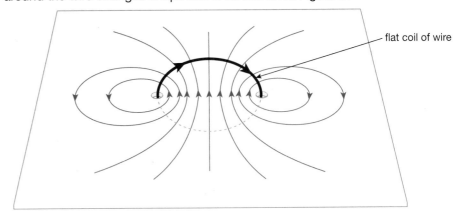

flat coil of wire

▲ Figure 20.11 The magnetic field around a flat coil

The strength of the magnetic field around a current-carrying wire can be increased by:

1 increasing the current in the wire

2 wrapping the wire into a coil or **solenoid** (a solenoid is a long coil).

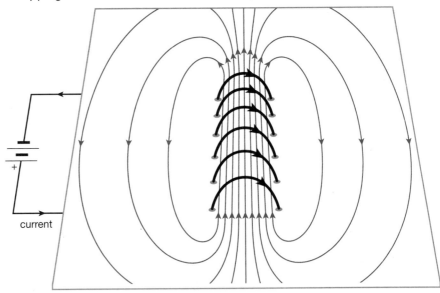

current

▲ Figure 20.12 The field around a solenoid looks like this.

The shape of the magnetic field around a solenoid is the same as that around a bar magnet.

EXTENSION WORK

You will not be asked questions about this but it is included in more advanced Physics courses.
The positions of the poles for a solenoid can be found using the right-hand grip rule (for poles).
If the direction of the current in the solenoid is reversed, so too are the positions of the poles.

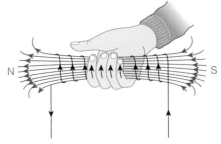

N   S

▲ Figure 20.13 You can work out the **polarity** of the solenoid by imagining that your right hand is wrapped around it. Your fingers point in the direction of the current and your thumb points to the north pole of the solenoid. This is the right-hand grip rule (for poles).

The strength of the field around a solenoid can be increased by:

1 increasing the current in the solenoid

2 increasing the number of turns on the solenoid

3 wrapping the solenoid around a magnetically soft core such as iron.

This combination of soft iron core and solenoid is often referred to as an electromagnet.

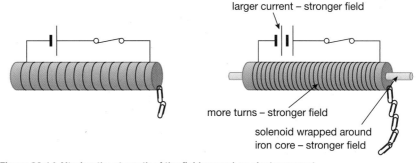

▲ Figure 20.14 Altering the strength of the field around an electromagnet

END OF PHYSICS ONLY

EXTENSION WORK

It is not necessary to memorise how an electric bell or a circuit breaker works but understanding how they work might help reinforce some of the basic principles of magnets and electromagnets you will need in your exams.

The electric bell

When the bell push is pressed the circuit is complete and there is current in the circuit. As a result the soft iron core of the electromagnet becomes magnetised and attracts the piece of iron that is attached to the hammer (this is called the armature). When the armature is pulled down, the hammer strikes the bell. At the same time a gap is created at the contact screw. The circuit is now incomplete and the current stops. The electromagnet is now turned off so the springy metal strip pulls the armature up to its original position.
The circuit is again complete and the whole process begins again.

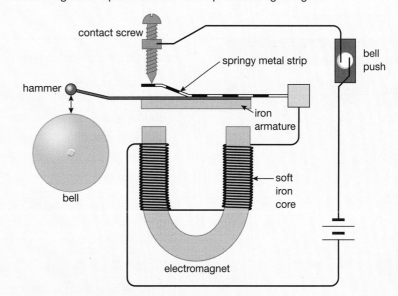

▲ Figure 20.15 An electric bell relies upon an electromagnet.

Circuit breaker

Circuit breakers also use electromagnets. The circuit breaker in Figure 20.16 uses electromagnetism to cut off the current if it becomes larger than a certain value. If the current is too high the electromagnet becomes strong enough to pull the iron catch out of position so that the contacts open and the circuit breaks. Once the problem in the circuit has been corrected the catch is repositioned by pressing the reset button.

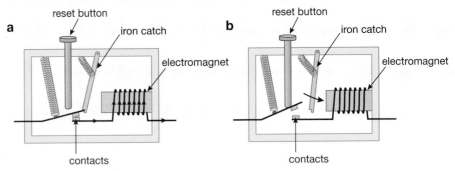

▲ Figure 20.16 **a** The circuit breaker allows a normal sized current. **b** When there is too large a current the circuit becomes incomplete.

▲ Figure 20.17 These circuit breakers control the supply of electricity to different circuits in a building. The circuit breaker will 'trip' if there is something wrong with the circuit and the current is too high.

CHAPTER QUESTIONS

More questions on electromagnetism can be found at the end of Unit 6 on page 217.

SKILLS CRITICAL THINKING

1 Which of these objects are made from magnetic materials?

a copper pipe   e iron bar

b plastic paper clip   f rubber tyre

c steel nail   g nickel coins

d glass bottle

SKILLS INTERPRETATION, REASONING

2 Draw a diagram of two bar magnets placed so that two of their poles are a repelling, and b attracting. Explain why you have drawn the magnets in these positions.

SKILLS CRITICAL THINKING

3 a What three features of a magnetic field do field lines show?

b What is a uniform magnetic field?

SKILLS INTERPRETATION

c Draw a diagram of a uniform field between two bar magnets.

4 a Draw a diagram to show the shape of the magnetic field around a solenoid.

b How does your diagram show where the magnetic field is strong or weak?

c Explain what happens to the magnetic field if the direction of the current through the solenoid is reversed.

5 In 1819 a scientist named Hans Christian Oersted was using a cell to produce a current in a wire. Close to the wire there was a compass. When there was a current in the wire, Oersted noticed – much to his surprise – that the compass needle moved.

a Why did the compass needle move?

b When there is a current in a horizontal wire, the needle of a compass placed beneath the wire comes to rest at right angles to the wire, pointing from left to right. In which direction will the compass needle point if it is held above the wire? Explain your answer.

c Would your answer to part b still be correct if the direction of the current in the wire was changed? Explain your answer.

6 The diagram below shows a simple electromagnet.

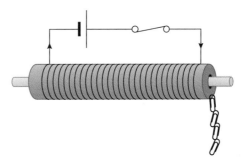

a Explain why the core (centre) of an electromagnet is made of soft iron and not steel.

b Suggest two ways in which the strength of this electromagnet could be increased.

21 ELECTRIC MOTORS AND ELECTROMAGNETIC INDUCTION

Trains in many cities use electric motors to transport millions of people to and from work each day. In the first part of this chapter, we are going to look at how an electric motor works. An electric motor creates motion from an electric current. Generators and alternators, on the other hand, transfer movement or kinetic energy into electrical energy by a process called electromagnetic induction. The second part of this chapter is going to look at this process.

▲ Figure 21.1 Electric train in Bangkok (the Skytrain)

LEARNING OBJECTIVES

PHYSICS ONLY

■ Know that there is a force on a charged particle when it moves in a magnetic field as long as its motion is not parallel to the field

■ Understand why a force is exerted on a current-carrying wire in a magnetic field, and how this effect is applied in simple d.c. electric motors and loudspeakers

■ Use the left-hand rule to predict the direction of the resulting force when a wire carries a current perpendicular to a magnetic field

■ Describe how the force on a current-carrying conductor in a magnetic field changes with the magnitude and direction of the field and current

■ Know that a voltage is induced in a conductor or a coil when it moves through a magnetic field or when a magnetic field changes through it and describe the factors that affect the size of the induced voltage

■ Describe the generation of electricity by the rotation of a magnet within a coil of wire and of a coil of wire within a magnetic field, and describe the factors that affect the size of the induced voltage

PHYSICS ONLY

■ Describe the structure of a transformer, and understand that a transformer changes the size of an alternating voltage by having different numbers of turns on the input and output sides

■ Explain the use of step-up and step-down transformers in the large-scale generation and transmission of electrical energy

■ Know and use the relationship between input (primary) and output (secondary) voltages and the turns ratio for a transformer:

$$\frac{\text{input (primary) voltage}}{\text{output (secondary) voltage}} = \frac{\text{primary turns}}{\text{secondary turns}}$$

■ Know and use the relationship:

input power = output power

$$V_p I_p = V_s I_s$$

for 100% efficiency

MOVEMENT FROM ELECTRICITY

When a charged particle moves through a magnetic field it experiences a force, as long as its motion is not parallel to the field. We can demonstrate this effect by passing electrons (that is, an electric current) along a wire that is placed in a magnetic field as shown in Figure 21.2. When the switch is closed and electrons flow, the wire will try to move upwards.

OVERLAPPING MAGNETIC FIELDS

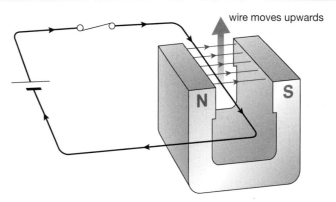

▲ Figure 21.2 The wire moves where there is a current in it.

We can explain this motion by looking at what happens when the switch is closed and the two magnetic fields overlap.

The two diagrams in Figure 21.3 show the shapes and directions of the magnetic fields **a** across the poles of the magnet, and **b** around the current-carrying wire before they overlap.

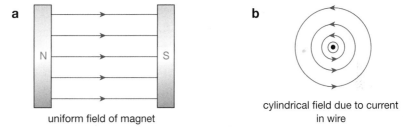

▲ Figure 21.3 **a** Magnet's uniform magnetic field **b** Magnetic field around a current carrying wire

If the wire is placed between the poles of the magnet, the two fields overlap.

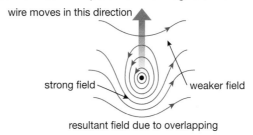

▲ Figure 21.4 The new magnetic field created by overlapping

In certain places, for example, below the wire, the fields are in the same direction and so reinforce each other. A strong magnetic field is produced here. In other places, for example, above the wire, the fields are in opposite directions. A weaker field is produced here. Because the fields are of different strengths the wire 'feels' a force, pushing it from the stronger part of the field to the weaker part – that is, in this case, upwards. The overlapping of the two magnetic fields has produced motion. This is called the motor effect.

A stronger force will be produced if the magnetic field is stronger or if the current is increased.

If we change the direction of the current or the magnet's field, a different overlapping pattern is created and we will see the wire move in the opposite direction.

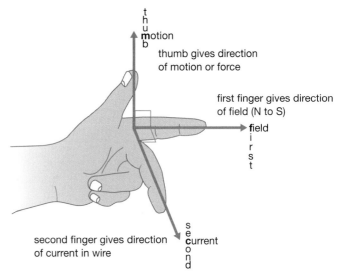

thumb
motion
thumb gives direction of motion or force

first finger gives direction of field (N to S)

field
first

second finger gives direction of current in wire

second
current

▲ Figure 21.5 Fleming's left-hand rule helps you to work out the direction of the force.

THE MOVING-COIL LOUDSPEAKER

The moving-coil loudspeaker uses the motor effect to transfer electrical energy to sound energy.

- Electric currents from a source, such as a radio, pass through the coils of a speaker.

- These currents, which represent sounds, are always changing in size and direction, like vibrating sound waves.

- The fields of the coil and the permanent magnet are therefore creating magnetic field patterns which are also always changing in strength and direction.

- These fields in turn apply rapidly changing forces to the wires of the coil, which cause the speaker cone to vibrate.

- These vibrations create the sound waves we hear.

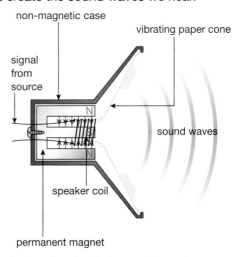

non-magnetic case

vibrating paper cone

signal from source

sound waves

speaker coil

permanent magnet

▲ Figure 21.6 A loudspeaker transfers electrical energy to sound energy.

THE ELECTRIC MOTOR

The electric motor is one of the most important uses of the motor effect. Figure 21.7 shows the most important features of a simple d.c. electric motor.

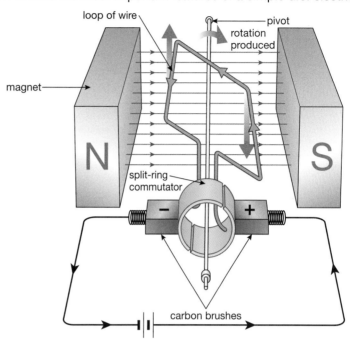

▲ Figure 21.7 A simple electric motor

When there is current in the loop of wire, one side of it will experience a force pushing it upwards. The other side will feel a force pushing it downwards, so the loop will begin to rotate (turn).

As the loop reaches the vertical position, its momentum takes it past the vertical. If the rotation is to continue the forces on the wires must now be reversed so that the wire at the top is now pushed down and the bottom one is pushed up.

This can be done easily by using a split ring to connect the loop of wire to the electrical supply. Now each time the loop of wire passes the vertical position, the connections change, the direction of the current changes, and the forces on the different sides of the loop change direction. The loop will rotate continuously.

To increase the rate at which the motor turns we can:

1 increase the number of turns or loops of wire, making a coil

2 increase the strength of the magnetic field

3 increase the current in the loop of wire.

EXTENSION WORK

Although you will not be asked this in your exam it is interesting to know how commercial motors like the one used in this electric drill differ from the simple motor described in Figure 21.7.

■ The permanent magnets are replaced with curved electromagnets capable of producing very strong magnetic fields.

■ The single loop of wire is replaced with several coils of wire wrapped on the same axis. This makes the motor more powerful and allows it to run more smoothly.

■ The coils are wrapped on a soft iron core that has been covered by a thin layer of plastic. This makes the motor more efficient and more powerful.

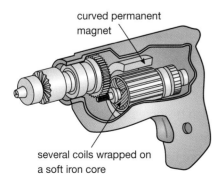

▲ Figure 21.8 A real electric motor

ELECTROMAGNETIC INDUCTION

Motors use electricity to produce movement. Generators and alternators are machines that use motion to produce electricity. They use a process called electromagnetic induction.

▲ Figure 21.9 Transportable generators are used to produce electricity for the lights and machinery used on roadworks.

The workers shown in Figure 21.9 need electricity for their machines and their lights. Instead of connecting into the mains supply, as we do at home, the workers have their own generator, which produces the electrical energy they need. In fact, the mains supply itself is produced by large generators in power stations. In this next section, you will discover how a generator produces electricity.

DEMONSTRATING ELECTROMAGNETIC INDUCTION

If we move a wire across a magnetic field at right angles, as shown in Figure 21.10, a voltage is induced or generated in the wire. If the wire is part of a complete circuit, there is a current. This event is called electromagnetic induction.

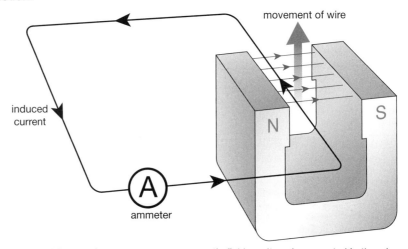

▲ Figure 21.10 When a wire moves across a magnetic field a voltage is generated in the wire.

The size of the induced voltage (and current) can be increased by:

1 moving the wire more quickly

2 using a stronger magnet so that there are more field lines 'cut'

3 wrapping the wire into a coil so that more pieces of wire move through the magnetic field.

We can also generate a voltage and current by pushing a magnet into a coil. In this situation and the previous one (Figure 21.10) we can see that it is the cutting action between the wires and the field lines that generates the voltage. If there is no cutting (that is, the wires and magnets are stationary) no voltage is generated.

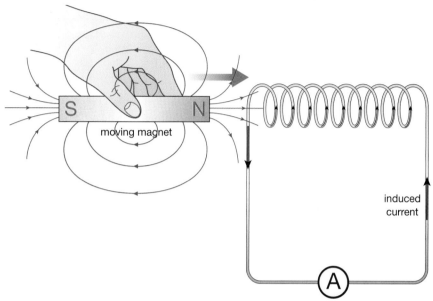

▲ Figure 21.11 A magnet moving in a coil will generate electricity.

This experiment also shows us that the size of the induced voltage (and current) can be increased by:

1 moving the magnet more quickly

2 using a stronger magnet

3 using a coil with more turns.

We can summarise all the discoveries from these experiments by saying:

■ a voltage is induced when a conductor cuts through magnetic field lines

■ a voltage is induced when magnetic field lines cut through a conductor

■ the faster the lines are cut the larger the induced voltage.

EXTENSION WORK

It is interesting to note, but not necessary for your exam, that Michael Faraday was the first person to observe how the size of an induced voltage depends upon the rate at which the magnetic lines of flux (field lines) are being cut. He summarised his observations in Faraday's Law of Electromagnetic Induction. This states that:

The size of the induced voltage across the ends of a wire (coil) is directly proportional to the rate at which the magnetic lines of flux are being cut.

GENERATORS

Figure 21.12 shows a small generator or dynamo used to generate electricity for a bicycle light.

As the cyclist pedals, the wheel rotates and makes a small magnet within the dynamo turn around. As this magnet turns, its magnetic field turns too. The field lines cut through the coil inducing a current in it. This current can be used to work the cyclist's lights.

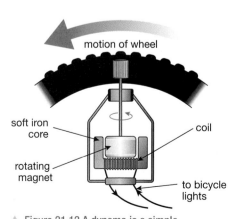

▲ Figure 21.12 A dynamo is a simple generator.

Figure 21.13 shows a much larger generator used in power stations to generate the mains electricity we use in our homes. In these generators the magnets are stationary and the coils rotate.

▲ Figure 21.13 This generator produces electricity on an industrial scale.

The size of the induced voltage is greater than in the bicycle dynamo because these generators **a** use much stronger magnets, **b** have many more turns of wire on the coil, and **c** spin the coil much faster.

PHYSICS ONLY

THE TRANSFORMER

When there is alternating current in a coil, the magnetic field around it is continuously changing. As the size of the current in the coil increases the field grows. As the size of the current decreases the field collapses. If a second coil is placed near the first, this growing and collapsing magnetic field will pass through it. As it cuts through the wires of the second coil, a voltage is induced across that coil. The size and direction of the induced voltage changes as the voltage applied to the first coil changes. (The first coil is more usually called the primary coil.) An alternating voltage applied across the primary coil therefore produces an alternating voltage across the secondary coil. This combination of two magnetically linked coils is called a transformer.

alternating input voltage, V_p — primary coil, with n_p turns — soft iron core linking the two coils — secondary coil, with n_s turns — alternating output voltage, V_s

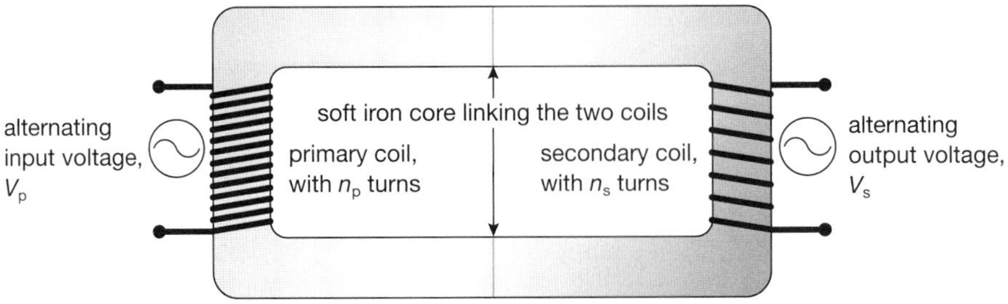

▲ Figure 21.14 The size of the voltage generated in the secondary coil of a transformer depends on the voltage in the primary coil and the numbers of turns on each coil.

REMINDER

Remember that transformers only work if the magnetic field around the primary coil is changing. Transformers will therefore only work with a.c. currents and voltages. They will not work with d.c. currents and voltages.

USING TRANSFORMERS TO CHANGE VOLTAGES

A **transformer** changes the size of an alternating voltage by having different numbers of turns on the input and output sides.

The relationship between the voltages across each of the coils is described by the equation:

$$\frac{\text{input voltage, } V_p}{\text{output voltage, } V_s} = \frac{\text{number of turns on primary coil, } n_p}{\text{number of turns on secondary coil, } n_s}$$

$$\frac{V_p}{V_s} = \frac{n_p}{n_s}$$

EXAMPLE 1

A transformer has 100 turns on its primary coil and 500 turns on its secondary coil. If an alternating voltage of 2 V is applied across the primary, what is the voltage across the secondary coil?

$$\frac{V_p}{V_s} = \frac{n_p}{n_s}$$

$$\frac{2\,V}{V_s} = \frac{100}{500}$$

$$V_s = \frac{500 \times 2\,V}{100}$$

$$= 10\,V$$

ENERGY IN TRANSFORMERS

It is very important to try to make transformers as efficient as possible. To keep any energy losses to a minimum we use thick copper wire for the coils and use a soft iron core covered in a thin layer of plastic to connect the two coils.

If a transformer is 100% efficient then the electrical energy entering the primary coil each second equals the electrical energy leaving the secondary coil each second: Another way of saying this is:

$$\text{input power, } V_p I_p = \text{output power, } V_s I_s$$

$$V_p I_p = V_s I_s$$

EXAMPLE 2

When a voltage of 12 V a.c. is applied across the primary coil of a step-down transformer, there is a current of 0.4 A in the primary coil. Calculate the current in the secondary coil if the voltage induced across it is 2 V a.c. Assume that the transformer is 100% efficient.

$$V_P I_P = V_S I_S$$

$$12\,V \times 0.4\,A = 2\,V \times I_S$$

$$I_S = \frac{12\,V \times 0.4\,A}{2\,V}$$

$$= 2.4\,A$$

▲ Figure 21.15 A transformer

TRANSFORMERS AND NATIONAL GRIDS

National Grids are networks of wires and cables that carry electrical energy from power stations to consumers such as factories and homes. Unfortunately, currents in long wires can lose lots of energy in the form of heat. The larger the current, the greater the amount of energy lost. If the current in the wires is kept to a minimum, the heat losses can be reduced. Transformers help us do this.

Transformers are used in National Grids so that the electricity is transmitted (sent out) as low currents and at high voltages.

Immediately after generation, electric currents from the alternators are passed through transformers which greatly decrease the size of the currents and increase the size of the voltages. These step-up transformers increase the voltage of the electricity to approximately 400 kV. High voltages like these can be extremely dangerous so the cables are supported high above the ground on pylons. As the cables enter towns and cities they are buried underground. Close to where the electrical energy is needed, the supply is passed through a step-down transformer that decreases the voltage to approximately 230 V, while at the same time increasing the current.

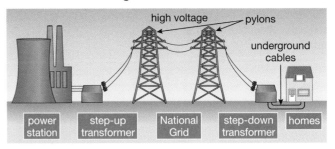

▲ Figure 21.16 Transformers are used in National Grids.

END OF PHYSICS ONLY

LOOKING AHEAD

Higher Physics courses are likely to ask you what exactly is happening inside a generator and what is an alternator. The simple description given below will give you some idea of what is required.

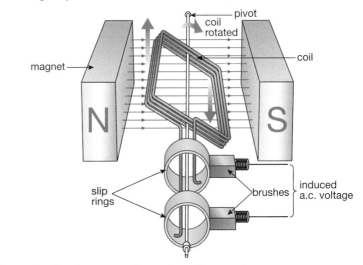

▲ Figure 21.17 The alternator produces alternating current.

As the coil rotates, its wires cut through magnetic field lines and a current is induced in them. If we watch just one side of the coil we see that the wire moves up through the field and then down for each turn of the coil. As a result the current is induced first in one direction and then in the opposite direction. This kind of current is called alternating current. A generator that produces alternating current is called an alternator.

The frequency of an alternating current is the number of complete cycles it makes each second. If an alternator coil rotates twice in a second, the frequency of the alternating current it produces is 2 Hz (2 cycles per second). The frequency of the UK mains supply is 50 Hz.

a

b
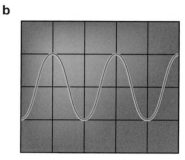

▲ Figure 21.18 **a** In a direct current (d.c) the charge flows in one direction. **b** Alternating currents (a.c.) keep changing direction.

CHAPTER QUESTIONS

More questions on electric motors and electromagnetic induction can be found at the end of Unit 6 on page 217.

More questions on electric motors and electromagnetic induction can be found at the end of Unit 6 on page 217.

SKILLS ANALYSIS, REASONING

1 The diagram below shows a long wire placed between the poles of a magnet.

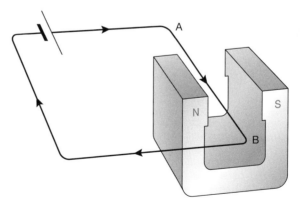

Describe what happens when:

a there is current in the wire from A to B

b the direction of the current is reversed – that is, from B to A

c with the current from B to A, the poles of the magnet are reversed

d there is a larger current in the wire.

SKILLS INTERPRETATION

2 a Draw a simple labelled diagram of a moving-coil loudspeaker.

SKILLS CRITICAL THINKING

b Explain how sound waves are made by the speaker when a signal is passed through its coil.

PHYSICS ONLY

3 a Explain the difference between a step-up transformer and a step-down transformer.

  b Explain where step-up and step-down transformers are used in National Grids.

  c Explain why transformers are used in National Grids.

  d Draw a fully labelled diagram of a transformer.

  e Explain why a transformer will not work if a d.c. voltage is applied across its primary coil.

  f A transformer has 200 turns on its primary coil and 5000 turns on its secondary coil. Calculate the voltage across the secondary coil when a voltage of 2 V a.c. is applied across the primary coil.

END OF PHYSICS ONLY

4 The diagram below shows a bar magnet being pushed into a long coil. A sensitive meter is connected across the coil.

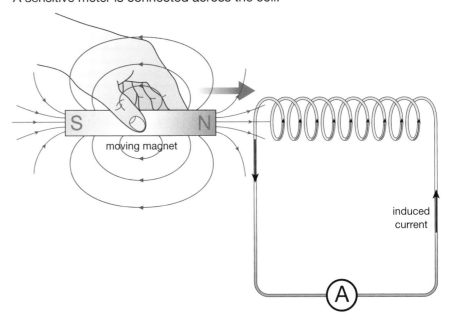

moving magnet

induced current

Describe what happens to the meter when:

a the magnet is pushed into the coil quickly

b the magnet is held stationary inside the coil

c the magnet is pulled out of the coil slowly

d the magnet is held stationary and the coil is moved towards it.

UNIT QUESTIONS

SKILLS ▶ CRITICAL THINKING

1 **a** Which of these objects is made from a magnetic material?

    A a rubber band

    B a piece of paper

    C a steel nail

    D a plastic ruler **(1)**

    b In which of these situations will you see attraction?

    A The north poles of two magnets are placed close together.

    B A wooden rod is placed close to a strong magnet.

    C The south poles of two magnets are placed close together.

    D The north pole of one magnet is placed close to the south pole of another magnet. **(1)**

    c Which of these statements is true for a magnetically hard material?

    A They are difficult to bend.

    B They are used to make permanent magnets.

    C They are electrical insulators.

    D They are used in electromagnets. **(1)**

(Total for Question 1 = 3 marks)

2 Copy and complete the following passage, filling in the blanks.

A wire carrying an electric current experiences a _____ when it is in a _____ field. This effect can be used in _____ and electric _____.

If a wire is moved through a magnetic field, or the magnetic field near a _____ of wire changes, a current can be _____ in the wire. This effect is used in _____ and _____.

(Total for Question 2 = 8 marks)

SKILLS ▶ REASONING

3 **a** Explain how a bicycle dynamo generates current. **(3)**

    b Why does the dynamo produce no current when the cyclist has stopped? **(1)**

SKILLS ▶ CRITICAL THINKING

    c What is an alternator? **(1)**

SKILLS ▶ INTERPRETATION, CRITICAL THINKING

    d Explain using diagrams the difference between a.c. and d.c. currents. **(3)**

SKILLS ▶ CRITICAL THINKING

    e Explain what is meant by this statement: 'The mains supply in the UK has a frequency of 50 Hz.' **(2)**

(Total for Question 3 = 10 marks)

PHYSICS ONLY

4

The diagram below shows how the electrical energy is transmitted from a power station to our homes.

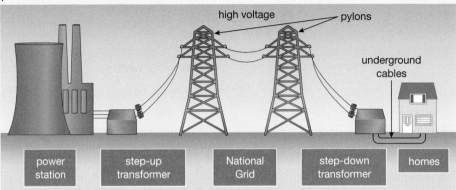

a Why is the voltage of the supply increased before transmission? **(1)**

b Why is the voltage of the supply decreased before entering our homes? **(1)**

c A step-up transformer has 100 turns on its primary coil and 20 000 turns on its secondary coil. Calculate the output voltage of the transformer if the input voltage is 12 V a.c. **(3)**

d Assuming that the transformer is 100% efficient, calculate the size of the induced current in the secondary coil if the current in the primary coil is 10 A. **(3)**

(Total for Question 4 = 8 marks)

END OF PHYSICS ONLY

5 A keen cyclist builds his own dynamo to generate electricity for his lights. Unfortunately, when he tries it out the lights are not bright enough. Suggest two changes he could make to his dynamo to generate more electricity. Is there a third solution to his problem?

(Total for Question 5 = 3 marks)

6 The diagram below shows the circuit for an electric bell.

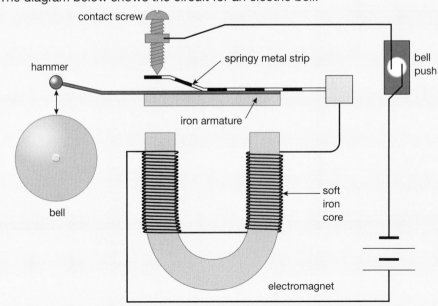

a Explain in your own words why the bell will not work if the electromagnet
   is replaced with a permanent magnet.   (2)

b Explain why the core of the electromagnet used in an electric bell must
   not be made from steel.   (3)

(Total for Question 6 = 5 marks)

PHYSICS ONLY

SKILLS REASONING

SKILLS PROBLEM SOLVING

a Explain the difference in the structure and use of a step-up transformer
   and a step-down transformer.   (4)

b A transformer has 2000 turns on its primary coil and 6000 turns on its
   secondary coil. If an a.c. voltage of 12 V is applied across the primary coil
   calculate the voltage induced across the secondary coil. Assume that the
   transformer is 100% efficient.   (4)

(Total for Question 7 = 8 marks)

END OF PHYSICS ONLY

UNIT 7 RADIOACTIVITY AND PARTICLES

Henri Becquerel was a scientist whose pioneering work led to the discovery of radioactivity and set other scientists on a path of research that continues today.

Using radioactivity has enabled us to learn about the structure of matter, to develop new sources of energy and to understand a new force of nature: nuclear forces that bind matter together and control the processes that fuel stars and formed all the elements that we and the Universe are made of.

22 ATOMS AND RADIOACTIVITY

Atoms are made up of sub-atomic particles called neutrons, protons and electrons. It is the numbers of these particles that give each element its unique properties. In this chapter you will find out how atoms can break up and become transformed into different elements, and about the different types of radiation they give out.

LEARNING OBJECTIVES

- Describe the structure of an atom in terms of protons, neutrons and electrons and use symbols such as $^{14}_{6}C$ to describe particular nuclei

- Know the terms atomic (proton) number, mass (nucleon) number and isotope

- Know that alpha (α) particles, beta (β⁻) particles, and gamma (γ) rays are ionising radiations emitted from unstable nuclei in a random process

- Describe the nature of alpha (α) particles, beta (β⁻) particles, and gamma (γ) rays, and recall that they may be distinguished in terms of penetrating power and ability to ionise

- Investigate the penetration powers of different types of radiation using either radioactive sources or simulations

- Describe the effects on the atomic and mass numbers of a nucleus of the emission of each of the four main types of radiation (alpha, beta, gamma and neutron radiation)

- Understand how to balance nuclear equations in terms of mass and charge

UNITS

In this section you will need to use becquerel (Bq) as the unit of activity of a radioactive source, centimetre (cm) as the unit of length, and minute (min) and second (s) as the units of time.

ELECTRONS, PROTONS AND NEUTRONS

Atoms are made up of electrons, protons and neutrons. Figure 22.1 shows a simple model of how these particles are arranged.

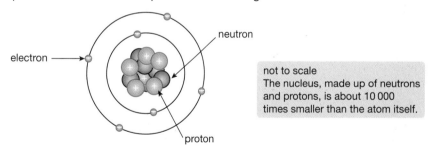

not to scale
The nucleus, made up of neutrons and protons, is about 10 000 times smaller than the atom itself.

▲ Figure 22.1 A simple model with protons and neutrons in the nucleus of the atom and electrons in orbits around the outside

The electron is a very small particle with very little mass. It has a negative electric charge. Electrons orbit the nucleus of the atom. The nucleus is very small compared to the size of the atom itself. The diameter of the nucleus is about 10 000 times smaller than the diameter of the atom. If the nucleus of an atom were enlarged to the size of a full stop on this page, the atom would have a diameter of around 2.5 metres.

The nucleus is made up of protons and neutrons. Protons and neutrons have almost exactly the same mass. Protons and neutrons are nearly 2000 times heavier than electrons. Protons carry positive electric charge but neutrons, as the name suggests, are electrically neutral or uncharged. The amount of charge on a proton is equal to that on an electron but opposite in sign.

The actual mass of an electron is 9.1×10^{-31} kg and its actual charge is -1.6×10^{-19} C. The table shows the approximate relative masses of the particles and the relative amount of charge they carry. The mass in atomic mass units is discussed below.

The properties of these three atomic particles are summarised in the table below. Protons and neutrons are also called nucleons because they are found in the nucleus of the atom.

Atomic particle	Relative mass of particle	Relative charge of particle
electron	1	−1
proton	2000	+1
neutron	2000	0

THE ATOM

The nucleus of an atom is surrounded by electrons. We sometimes think of electrons as orbiting the nucleus in a way similar to the planets orbiting the Sun. It is more accurate to think of the electrons as moving rapidly around the nucleus in a cloud or shell.

An atom is electrically neutral. This is because the number of positive charges carried by the protons in its nucleus is balanced by the number of negative charges on the electrons in the electron 'cloud' around the nucleus.

ATOMIC NUMBER, Z

The chemical behaviour and properties of a particular element depend upon how the atoms combine with other atoms. This is determined by the number of electrons in the atom. Although atoms may gain or lose electrons, sometimes quite easily, the number of protons in atoms of a particular element is always the same. The atomic number of an element tells us how many protons each of its atoms contains. For example, carbon has six protons in its nucleus – the atomic number of carbon is, therefore, 6. The symbol we use for atomic number is Z. Each element has its own unique atomic number. The atomic number is sometimes called the proton number.

ATOMIC MASS, A

The total number of protons and neutrons in the nucleus of an atom determines its atomic mass. The mass of the electrons that make up an atom is tiny and can usually be ignored. The mass of a proton is approximately 1.7×10^{-27} kg. To save writing this down we usually refer to the mass of an atom by its mass number or nucleon number. This number is the total number of protons and neutrons in the atom. The mass number of an element is given the symbol A.

ATOMIC NOTATION – THE RECIPE FOR AN ATOM

Each particular type of atom will have its own atomic number, which identifies the element, and a mass number that depends on the total number of nucleons, or particles, in the nucleus. Figure 22.2 shows the way we represent an atom of an element whose chemical symbol is X, showing the atomic number and the mass number.

So, using this notation, an atom of oxygen is represented by:

$$^{16}_{8}O$$

The chemical symbol for oxygen is O. The atomic number is 8 – this tells us that the nucleus contains eight protons. The mass number is 16, so there are 16 nucleons (protons and neutrons) in the nucleus. Since eight of these are protons, the remaining eight must be neutrons. The atom is electrically neutral overall, so the +8 charge of the nucleus is balanced by the eight orbiting electrons, each with charge −1.

KEY POINT

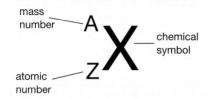

▲ Figure 22.2 Atomic notation

mass number, A = number of neutrons + number of protons, Z = number of nucleons

so

number of neutrons = number of nucleons – number of protons = A – Z

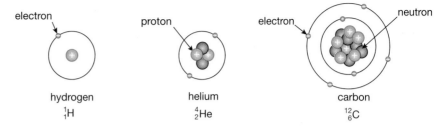

▲ Figure 22.3 The hydrogen atom has one proton in its nucleus and no neutrons, so the mass number A = 1 + 0 = 1. As it has one proton, its atomic number, Z = 1. For helium, A = 4 (2 protons + 2 neutrons) and Z = 2 (2 protons). For carbon, A = 12 (6 protons + 6 neutrons) while Z = 6 (6 protons).

Figure 22.3 shows some examples of the use of this notation for hydrogen, helium and carbon, together with a simple indication of the structure of an atom of each of these elements. In each case the number of orbiting electrons is equal to the number of protons in the nucleus, so the atoms are electrically neutral.

ISOTOPES

The number of protons in an atom identifies the element. The chemical behaviour of an element depends on the number of electrons it has and, as we have seen, this always balances the number of protons in the nucleus. However, the number of neutrons in the nucleus can vary slightly. Atoms of an element with different numbers of neutrons are called isotopes of the element. The number of neutrons in a nucleus affects the mass of the atom. Different isotopes of an element will all have the same atomic number, Z, but different mass numbers, A. Figures 22.4 and 22.5 show some examples of isotopes.

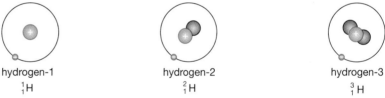

▲ Figure 22.4 Isotopes of hydrogen – they all have the same atomic number, 1, and the same chemical symbol, H.

EXTENSION WORK

Hydrogen-2 is also called heavy hydrogen or deuterium. Hydrogen-3 is called tritium.

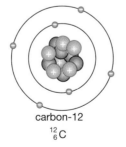

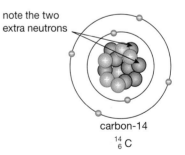

▲ Figure 22.5 Two isotopes of carbon – they are referred to as carbon-12 and carbon-14 to distinguish between them.

THE STABILITY OF ISOTOPES

Isotopes of an element have different physical properties from other isotopes of the same element. One obvious difference is the mass. Another difference is the stability of the nucleus.

The protons are held in the nucleus by the nuclear force. This force is very strong and acts over a very small distance. It is strong enough to hold the nucleus together against the electric force repelling the protons away from each other. (Remember that protons carry positive charge and like charges repel.) The presence of neutrons in the nucleus affects the balance between these forces. Too many or too few neutrons will make the nucleus unstable. An unstable nucleus will eventually decay. When the nucleus of an atom decays it gives out energy and may also give out alpha or beta particles.

IONISING RADIATION

When unstable nuclei decay they give out ionising radiation. Ionising radiation causes atoms to gain or lose electric charge, forming ions. Unstable nuclei decay at random. This means that it is not possible to predict which unstable nucleus in a piece of radioactive material will decay, or when decay will happen. We shall see that we can make measurements that will enable us to predict the probability that a certain proportion of a radioactive material will decay in a given time.

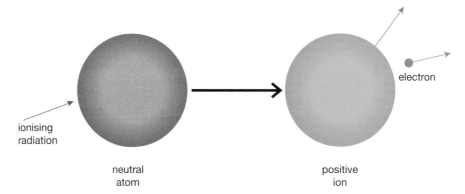

ionising radiation

neutral atom

positive ion

electron

▲ Figure 22.6 When a neutral atom (or molecule) is hit by ionising radiation it loses an electron and becomes a positively charged ion.

There are three basic types of ionising radiation. They are: alpha (α), beta (β) and gamma (γ) radiation.

ALPHA (α) RADIATION

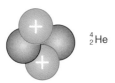

$^{4}_{2}He$

▲ Figure 22.7 An alpha particle (α particle)

EXTENSION WORK

Nucleons have roughly 2000 times the mass of an electron, and alpha particles are made up of four nucleons, so an alpha particle has 8000 times the mass of a beta particle.

Alpha radiation consists of fast-moving particles that are thrown out of unstable nuclei when they decay. These are called alpha particles. Alpha particles are helium nuclei – helium atoms without their orbiting electrons. Figure 22.7 shows an alpha particle and the notation that is used to denote it in equations.

Alpha particles have a relatively large mass. They are made up of four nucleons and so have a mass number of 4. They are also charged because of the two protons that they carry. The relative charge of an alpha particle is +2.

Alpha particles have a short range. The range of ionising radiation is the distance it can travel through matter. Alpha particles can only travel a few centimetres in air and cannot penetrate more than a few millimetres of paper. They have a limited range because they interact with atoms along their paths, causing ions to form. This means that they rapidly give up the energy that they had when they were ejected from the unstable nucleus.

BETA RADIATION (β⁻ AND β⁺)

Beta minus particles (β⁻) are very fast-moving electrons that are ejected by a decaying nucleus. The nucleus of an atom contains protons and neutrons, so where does the electron come from? The stability of a nucleus depends on the proportion of protons and neutrons it contains. The result of radioactive decay is to change the balance of protons and neutrons in the nucleus to make it

EXTENSION WORK

Antimatter will not be examined, but for all the particles that exist in ordinary matter there exists an antimatter particle. The positron is the antimatter equivalent of an electron. You will not be asked about neutrinos or antineutrinos either.

more stable. Beta minus decay involves a neutron in the nucleus splitting into a proton and an electron (an antineutrino is also emitted). The proton remains in the nucleus and the electron is ejected at high speed as a beta minus particle.

EXTENSION WORK

Sometimes a proton will decay into a neutron by emitting a beta plus particle (β^+). A beta plus particle is an antielectron, a particle with the same mass as an electron but with positive charge. The antielectron is called a positron.

Beta particles are very light – they have only 0.000 125 times the mass of an alpha particle. The relative charge of a β^- is –1 and the relative charge of a β^+ is +1.

Beta particles interact with matter in their paths less frequently than alpha particles. This is because they are smaller and carry less charge. This means that beta particles have a greater range than alpha particles. Beta particles can travel long distances through air, pass through paper easily and are only absorbed by denser materials like aluminium. A millimetre or two of aluminium foil will stop all but the most energetic beta particles.

GAMMA RAYS (γ)

EXTENSION WORK

Gamma radiation is emitted in 'packets' of energy called photons.

Gamma rays are electromagnetic waves (see page 106) with very short wavelengths. As they are waves, they have no mass and no charge. They are weakly ionising and interact only occasionally with atoms in their paths. They are extremely penetrating and pass through all but the very densest materials with ease. It takes several centimetres thickness of lead, or a metre or so of concrete, to stop gamma radiation.

NEUTRON RADIATION

Neutrons are emitted by radioactive material. They have roughly the same mass as a proton but have no electric charge. The symbol used for a neutron in radioactive decay equations is:

$$^1_0 n$$

If a neutron is emitted by a nucleus its mass number (A) goes down by 1 but the atomic number (Z) is unchanged. If a neutron is absorbed by a nucleus its mass number (A) goes up by 1 but the atomic number (Z) is unchanged.

EXTENSION WORK

As neutrons cause other atoms to emit radiation they must be screened. This is done using materials that have a lot of light atoms like hydrogen. Concrete and water are used to prevent neutrons escaping from nuclear reactors.

KEY POINT

As neutrons are not electrically charged they do not directly cause ionisation. They are absorbed by nuclei of other atoms and can cause them to become radioactive. The radioactive nuclei formed in this way will then decay emitting ionising radiation. Neutrons are the only type of radiation to cause other atoms to become radioactive. We shall see later (Chapter 25) that neutron radiation plays an important part in the process of fission used in nuclear reactors.

SUMMARY OF THE PROPERTIES OF IONISING RADIATION

We have said that ionising radiation causes uncharged atoms to lose electrons. An atom that has lost (or gained) electrons has an overall charge. It is called an ion. The three types of radioactive emission can all form ions.

As ionising radiation passes through matter, its energy is absorbed. This means that radiation can only penetrate matter up to a certain thickness. This depends on the type of radiation and the density of the material that it is passing through.

The ionising and penetrating powers of alpha, beta and gamma radiation are compared in the table on page 226. Note that the ranges given in the table are typical but they do depend on the energy of the radiation. For example, more energetic alpha particles will have a greater range than those with lower energy.

Radiation	Ionising power	Penetrating power	Example of range in air	Radiation stopped by
alpha, α	strong	weak	5–8 cm	paper
beta, β	medium	medium	500–1000 cm	thin aluminium
gamma, γ	weak	strong	virtually infinite	thick lead sheet

ACTIVITY 1

▼ PRACTICAL: INVESTIGATE THE PENETRATING POWERS OF DIFFERENT KINDS OF RADIATION

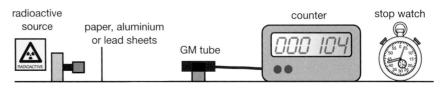

▲ Figure 22.8 Measuring the penetrating power of different types of radiation

Radioactive sources must be stored in lead-lined boxes and kept in a metal cupboard with a radiation warning label. The source must be handled with tongs away from the body. (See Chapter 24 page 247 for more details of safe practice with radioactive materials.)

Before the source is removed from its storage container, measure the background radiation count by connecting a Geiger–Müller (GM) tube to a counter. Write down the number of counts after 5 minutes. Repeat this three times and find the average background radiation count.

Take a source of alpha radiation and set it up at a measured distance (between 2 and 4 cm) from the GM tube.

Measure the counts detected in a 5 minute period. Repeat the count with a sheet of thick paper in front of the source. You should find that the counts have dropped to the background radiation count. This shows that alpha radiation does not pass through paper.

Now replace the alpha source with a beta source. After measuring the new count for 5 minutes place thin sheets of aluminium between the source and detector. When the thickness of the aluminium sheet is 1–2 mm thick you will find that the count has dropped to the background radiation level. This shows that beta radiation is blocked by just a few millimetres of aluminium.

Finally carry out the same steps using a gamma radiation source. Now you will find that gamma radiation is only blocked when a few centimetres of lead are placed between the source and the detector.

KEY POINT

The Geiger–Müller (GM) tube is a detector of radiation. It is described in the next chapter.

As we said earlier, an unstable atom, or strictly speaking its nucleus, will decay by emitting radiation. If the decay process involves the nucleus ejecting either an alpha or a beta particle, the atomic number will change. This means that alpha or beta decay causes the original element to transform into a different element.

ALPHA (α) DECAY

Here is an example of alpha decay:

$$^{222}_{88}\text{Ra} \rightarrow ^{218}_{86}\text{Ra} + ^{4}_{2}\text{He} + \text{energy}$$

radium atom → radon atom + α particle + energy

The radioactive isotope radium-222 decays to the element radon by the emission of an alpha particle. The alpha particle is sometimes represented by the Greek letter α. Radon is a radioactive gas that also decays by emitting an alpha particle. Note that the atomic number for radon, 86, is two less than the atomic number for radium.

This is a balanced equation:

■ The total of the A numbers on each side of the equation is the same. (Remember that A is the total number of protons and neutrons.)

■ The total of the Z numbers on each side of the equation is the same. (Remember that the Z number tells us the number of protons in the nucleus – the number of positively charged particles.)

The general form of the alpha decay equation is:

$$^{A}_{Z}\text{Y} \rightarrow ^{A-4}_{Z-2}\text{W} + ^{4}_{2}\text{He} + \text{energy}$$

alpha particle, α

In alpha decay, element Y is transformed into element W by the emission of an alpha particle. Element W is two places before element Y in the periodic table. The alpha particle, a helium nucleus, carries away four nucleons, which reduces the mass number (A) by four. Two of these nucleons are protons so the atomic number of the new element is two less than the original element, Z – 2. Notice that the mass number and the atomic number are conserved through this equation – that is, the total numbers of nucleons and protons on each side of the equation are the same.

REMINDER

A is the mass number of the element and Z is the atomic number. The letters W and Y are not the symbols of any particular elements as this is the general equation.

KEY POINT

It is worth pointing out that the mass number refers to the number of nuclear particles, or nucleons, involved in the transformation – not the exact mass. Mass is not conserved in nuclear transformations, as some of it is transformed into energy.

EXAMPLE 1

Complete the following nuclear equation which shows the decay of uranium-238 to the element thorium by alpha decay (emitting an alpha particle).

$$^{238}_{92}\text{U} \rightarrow ^{?}_{?}\text{Th} \rightarrow ^{4}_{2}\text{He}$$

The atomic mass numbers (A) on both sides of the equation must add up to the same, so the atomic mass for thorium must be 234 (238 = 234 + 4).

The atomic numbers (Z) on both sides of the equation must add up to the same, so the atomic mass for thorium must be 90 (92 = 90 + 2). So the answer is:

$$^{238}_{92}\text{U} \rightarrow ^{234}_{90}\text{Th} \rightarrow ^{4}_{2}\text{He}$$

KEY POINT

A complete periodic table is shown in Appendix A on page 277. The numbers shown in the top left corner are the atomic numbers of the elements – the Z numbers in the nuclear equations. Remember that the Z numbers tell us the number of protons in the nucleus of an element. Some periodic tables show the atomic mass too but as elements have a number of different isotopes the mass numbers will be an average for the different isotopes.

Figure 22.9 shows the effect of alpha particle decay. Losing an alpha particle from the nucleus of a uranium atom turns it into thorium, two places back in the periodic table.

Ac	Th	Pa	U	Np
Actinium	Thorium	Protactinium	Uranium	Neptunium
89	90	91	92	93

▲ Figure 22.9 A part of the periodic table of elements

Here is an example of beta minus decay:

$$^{14}_{6}\text{C} \quad \rightarrow \quad ^{14}_{7}\text{N} \quad + \quad ^{0}_{-1}\text{e} \quad + \text{energy}$$

carbon-14 atom → nitrogen-14 atom + beta particle, β⁻ + energy (also an antineutrino)

The radioactive isotope of carbon, carbon-14, decays to form the stable isotope of the gas nitrogen, by emitting a beta particle. Remember that the beta minus particle is formed when a neutron splits to form a proton and an electron. Figure 22.10 shows the standard atomic notation for a beta particle (β⁻).

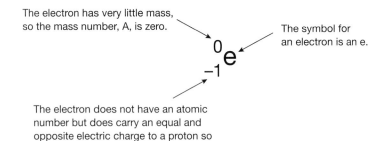

The electron has very little mass, so the mass number, A, is zero.

The symbol for an electron is an e.

The electron does not have an atomic number but does carry an equal and opposite electric charge to a proton so its Z number is –1.

▲ Figure 22.10 A beta minus particle

The general form of the beta minus decay equation is:

$$^{A}_{Z}\text{X} \rightarrow ^{A}_{Z+1}\text{Y} + ^{0}_{-1}\text{e} + \text{energy}$$

beta minus particle, β⁻

In β⁻ decay, element X is transformed into element Y by the emission of a beta minus particle. Element Y is the next element in the periodic table after element X. The beta particle, an electron, has practically no mass so the mass number, A, is the same in X and Y. As the β⁻ particle has a charge of –1, the atomic number of the new element is increased to Z + 1. Again the mass number and the atomic number are conserved through this equation.

EXTENSION WORK

You will not be asked about β⁺ decay but the example here is included to show that the atomic mass numbers (A) and the atomic numbers (Z) on each side of the decay equation must balance. Here is an example of beta plus decay:

$$^{22}_{11}\text{Na} \quad \rightarrow \quad ^{22}_{10}\text{Ne} \quad + \quad ^{0}_{+1}\text{e} \quad + \text{energy}$$

sodium-22 atom → neon-22 atom + beta particle, β⁺ + energy (also a neutrino)

The radioactive isotope of sodium (Na), sodium-22, decays to form the stable isotope of the gas neon (Ne), by emitting a beta plus particle. Remember that the beta plus particle is formed when a proton splits to form a neutron and a positron. Figure 22.11 shows the standard atomic notation for a beta particle (β⁺).

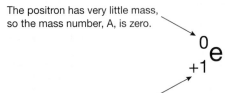

The positron has very little mass, so the mass number, A, is zero.

The electron does not have an atomic number but has the same electric charge as a proton so its Z number is +1.

▲ Figure 22.11 A beta plus particle

The general form of the beta plus decay equation is:

$$^{A}_{Z}X \rightarrow\ ^{A}_{Z-1}Y\ +\ ^{0}_{+1}e\ +\ \text{energy}$$

beta plus particle, β^{+}

In β^{+} decay, element X is transformed into element Y by the emission of a beta plus particle. Element Y is the element in the periodic table that comes before element X. The beta plus particle, a positron, has practically no mass so the mass number, A, is the same in X and Y. As this beta particle has a charge of +1, the atomic number of the new element is decreased to Z – 1.

EXAMPLE 2

An isotope of caesium $^{137}_{55}$Cs decays by emitting an electron (β^{-}) and forms an atom of barium (Ba). Write a balanced nuclear equation for this decay.

$$^{137}_{55}\text{Cs} \rightarrow\ ^{?}_{?}\text{Ba} +\ ^{0}_{-1}e + \text{energy}$$

An electron has practically no mass so the mass number does not change. The emission of an electron happens when a neutron becomes a proton and an electron, so the atomic number increases by one. The barium isotope formed is therefore $^{137}_{56}$Ba and the completed equation is:

$$^{137}_{55}\text{Cs} \rightarrow\ ^{137}_{56}\text{Ba} +\ ^{0}_{-1}e + \text{energy}$$

When an element decays by emitting an electron it changes to the element with the next higher atomic number. You can check this by looking at the periodic table on page 277.

REMINDER

You will not be asked questions involving positrons in the exam.

EXAMPLE 3

An isotope of magnesium $^{23}_{12}$Mg decays by emitting a positron (β^{+}) and forms an atom of sodium (Na). Write a balanced nuclear equation for this decay.

$$^{23}_{12}\text{Mg} \rightarrow\ ^{?}_{?}\text{Na} +\ ^{0}_{+1}e$$

A positron has practically no mass so the mass number does not change. The emission of a positron happens when a proton in the nucleus becomes a neutron and a positron, so the atomic number decreases by one. The sodium isotope formed is therefore $^{23}_{11}$Na and the completed equation is:

$$^{23}_{12}\text{Cs} \rightarrow\ ^{23}_{11}\text{Na} +\ ^{0}_{+1}e$$

When an element decays by emitting a positron it changes to the element with the next lower atomic number. You can check this by looking at the periodic table on page 277.

NEUTRON DECAY

Here is an example of neutron decay:

$$^{5}_{2}\text{He} \quad \rightarrow \quad ^{4}_{2}\text{He} \quad + \quad ^{1}_{0}n$$

helium-5 atom → helium-4 atom + neutron

As a neutron has no electric charge, the total positive charge is unchanged by the emission of a neutron – the Z number is 2 before and after the decay. The total number of particles (protons and neutrons) in the nucleus of the helium atom has decreased by 1 because of emission of the neutron.

EXAMPLE 4

An isotope of beryllium, $^{13}_{4}\text{Be}$, decays by emitting a neutron. Write a balanced nuclear equation for this transformation.

Write down what you know:

$$^{13}_{4}\text{Be} \rightarrow {}^{?}_{?}\text{?} + {}^{1}_{0}\text{n}$$

The emission of a neutron has no effect on the charge so the atomic number (Z) is unchanged at 4. Each element has its own atomic number so the element in this decay is unchanged – it is still an isotope of beryllium. The loss of a neutron from the beryllium nucleus means that the atomic mass decreases by one. The completed equation is:

$$^{13}_{4}\text{Be} \rightarrow {}^{12}_{4}\text{Be} + {}^{1}_{0}\text{n}$$

GAMMA (γ) DECAY

Gamma radiation is high-energy electromagnetic radiation (see page 106). After an unstable nucleus has emitted an alpha or beta particle it sometimes has surplus energy. It emits this energy as gamma radiation. Gamma rays are pure energy, so they do not have any mass or charge. When a nucleus emits a gamma ray there is no change to either the atomic number or the mass number of the nucleus.

LOOKING AHEAD

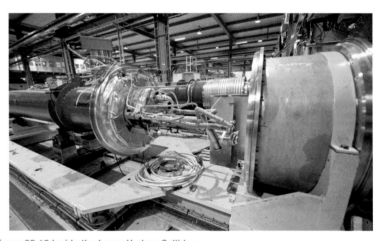

▲ Figure 22.12 Inside the Large Hadron Collider

Understanding the particles that make up all matter in the Universe, how they interact and the forces involved holding matter together remains a major area of research for physicists. This is an introduction to what has been called the 'particle zoo'. Large particle accelerators, like the Large Hadron Collider (LHC) on the Swiss/French border, have identified many new particles. Energy turned into matter! Research has recognised patterns in these particles and we now have a model that explains many complex nuclear processes – some of which have increased our understanding of how the Universe was formed.

We also understand how matter can be transformed into energy and the amazing amount of energy that results, revealed by Albert Einstein in his famous equation:

$$E = mc^2$$

CHAPTER QUESTIONS

SKILLS CRITICAL THINKING

More questions on the structure of the atom can be found at the end of Unit 7 on page 255.

1 Copy and complete the table below. Identify the particles and complete the missing data.

Atomic particle	Relative mass of particle	Relative charge of particle
	1	−1
		+1
	2000	0

2 Identify the following atomic particles from their descriptions:
a an uncharged nucleon
b the particle with the least mass
c the particle with the same mass as a neutron
d the particle with the same amount of charge as an electron
e a particle that is negatively charged.

3 Explain the following terms used to describe the structure of an atom:
a atomic number
b mass number.

4 Copy and complete the table below, describing the structures of the different atoms in terms of numbers of protons, neutrons and electrons.

	$^{3}_{2}\text{He}$	$^{13}_{6}\text{C}$	$^{23}_{11}\text{N}$
protons			
neutrons			
electrons			

5 Copy and complete the following sentences:
a An alpha particle consists of four _____. Two of these are _____ and two are _____. An alpha particle carries a charge of _____.
b A beta minus particle is a fast-moving _____ that is emitted from the nucleus. It is created when a _____ in the nucleus decays to form a _____ and the beta particle.
c A third type of ionising radiation has no mass. It is called _____ radiation. This type of radiation is a type of wave with a very _____ wavelength.
d Gamma radiation is part of the _____ spectrum.

SKILLS PROBLEM SOLVING

6 The following nuclear equation showing beta plus decay:
$$^{1}_{1}\text{p} \rightarrow {}^{1}_{0}\square + {}^{\square}_{\square}\text{e}$$

a From the following list of particles:
A electron   B neutron   C proton   D alpha particle
i State which is the particle on the left of this balanced equation.
ii State which is the first of the terms on the right of this equation.
b The second term on the right of this equation is a positron – the antiparticle of an electron. Copy and complete the above equation filling in the empty boxes shown.

HINT

Look at the table on page 226.

7 A certain radioactive source emits different types of radiation. The sample is tested using a Geiger counter. When a piece of card is placed between the source and the counter, there is a noticeable drop in the radiation. When a thin sheet of aluminium is added to the card between the source and the counter, the count rate is unchanged. A thick block of lead, however, causes the count to fall to the background level.

What type (or types) of ionising radiation is the source emitting? Explain your answer carefully.

8 a The nuclear equation below shows the decay of thorium. Copy and complete the equation by providing the missing numbers.

$$^{234}_{90}\text{Th} \rightarrow {}^{\square}_{\square}\text{Pa} + {}^{0}_{-1}\text{e}$$

   b What type of decay is taking place in this transformation?

   c The nuclear equation below shows the decay of polonium. Copy and complete the equation by providing the missing numbers and letters.

$$^{216}_{\square}\text{Po} \rightarrow {}^{\square}_{82}\text{Pb} + {}^{4}_{2}\square$$

   d What type of decay is taking place in this transformation?

23 RADIATION AND HALF-LIFE

In this chapter you will learn about ways of detecting radiation, and where some of the radiation around us comes from. You will also learn why we use a value called half-life to describe the activity of radioactive isotopes.

LEARNING OBJECTIVES

- Know that photographic film or a Geiger–Müller detector can detect ionising radiation

- Explain the sources of background (ionising) radiation from Earth and space

- Know that the activity of a radioactive source decreases over time and is measured in becquerels

- Know the definition of the term half-life and understand that it is different for different radioactive isotopes

- Use the concept of the half-life to carry out simple calculations on activity, including graphical methods

DETECTING IONISING RADIATION

USING PHOTOGRAPHIC FILM

Wilhelm Röntgen (Figure 23.1a) discovered x-rays in 1895. Henri Becquerel (Figure 23.1b) believed that uranium emitted x-rays after being exposed to sunlight. To test this idea he placed some wrapped, unused photographic plates in a drawer with some samples of uranium ore on top of them. He found a strong image of the ore on the plates when he developed them. He later realised that this was due to a new type of ionising radiation. He had discovered radioactivity.

The unit of radioactivity is named after Becquerel. The becquerel (Bq) is a measure of how many unstable nuclei are disintegrating (breaking up) per second – one becquerel means a rate of one disintegration per second. The becquerel is a very small unit. More practical units are the kBq (an average of 1000 disintegrations per second) and the MBq (an average of 1 000 000 disintegrations per second).

Photographic film can still be used to detect radioactivity. Some scientists who work with radioactive materials wear a piece of photographic film in a badge. If the film becomes fogged (unclear) it shows that the scientist has been exposed to a certain amount of radiation. These badges are checked regularly to ensure that scientists are not exposed to too much ionising radiation.

a   b

 Figure 23.1 **a** Wilhelm Röntgen (1845–1923), **b** Henri Becquerel (1852–1908). Nobel prize winners in 1901 and 1903

Figure 23.2 shows the basic construction of a Geiger–Müller (GM) tube. It is a glass tube with an electrically conducting surface on the inside. The tube has a thin window made of mica (a naturally occurring mineral that can be split into thin sheets). The tube contains a special mixture of gases at very low pressure. In the middle of the tube, electrically insulated from the conducting surface, there is an **electrode**. This electrode is connected, via a high-value resistor, to a high-voltage supply, typically 300–500 V.

When ionising radiation enters the tube it causes the low pressure gas inside to form ions. The ions allow a small amount of current to flow from the electrode to the conducting layer. This is detected by an electronic circuit.

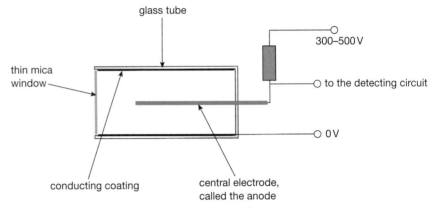

▲ Figure 23.2 A Geiger–Müller tube is used to measure the level of radiation.

The GM tube is usually linked up to a counting circuit. This keeps a count of how many ionising particles (or how much γ radiation) have entered the GM tube. Sometimes GM tubes are connected to rate meters. These measure the number of ionising events per second, and so give a measure of the radioactivity in becquerels. Rate meters usually have a loudspeaker output so the level of radioactivity is indicated by the rate of clicks produced.

BACKGROUND RADIATION

Background radiation is low-level ionising radiation that is produced all the time. This background radiation has a number of sources. Some of these are natural and some are artificial.

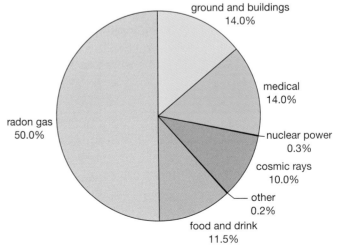

▲ Figure 23.3 Sources of background radiation in the UK. These are the average values – the true amounts and proportions vary from place to place.

NATURAL BACKGROUND RADIATION FROM THE EARTH

KEY POINT

When an atom of a radioactive element decays it gives out radiation and changes to an atom of another element. This may also be radioactive, and decay to form an atom of yet another element. The elements formed as a result of a radioactive element experiencing a series of decays are called decay products.

Some of the radiation we receive comes from rocks in the Earth's crust (hard outer layer). When the Earth was formed, around 4.5 billion years ago, it contained many radioactive isotopes. Some decayed very quickly but others are still producing radiation. Some of the decay products of these long-lived radioactive materials are also radioactive, so there are radioactive isotopes with much shorter half-lives (see page 237) still present in the Earth's crust.

One form of uranium is a radioactive element that decays very slowly. Two of its decay products are gases. These are the radioactive gases radon and thoron. Radon-222 is a highly radioactive gas produced by the decay of radium-226. Thoron, or radium-220, is an isotope of radium formed by the decay of a radioactive isotope of thorium (thorium-232).

As these decay products are gases, they come out of radioactive rocks. They are dense gases so they build up in the basements of buildings. Some parts of the Earth's crust have higher amounts of radioactive material so the amount of background radiation produced in this way varies from place to place. In Cornwall in the UK, for example, where the granite rock contains traces of uranium, the risk of exposure to radiation from radon gas is greater than in some other parts of the UK.

NATURAL BACKGROUND RADIATION FROM SPACE

Violent nuclear reactions in stars and exploding stars called supernovae produce cosmic rays (very energetic particles) that continuously hit the Earth. Lower energy cosmic rays are given out by the Sun. Our atmosphere gives us fairly good protection from cosmic rays but some still reach the Earth's surface.

RADIATION IN LIVING THINGS

The atoms that make up our bodies were formed in the violent reactions that take place in stars that exploded (supernovas) billions of years ago. Some of these atoms are radioactive so we carry our own personal source of radiation around with us. Also, as we breathe we take in tiny amounts of the radioactive isotope of carbon, carbon-14. Because carbon-14 behaves chemically just like the stable isotope, carbon-12, we continuously renew the amount of the radioactive carbon in our bodies (see page 245).

Carbon-14 and other radioactive isotopes are eaten by humans (and animals which are in turn eaten by humans) because they are present in all living things.

ARTIFICIAL RADIATION

We use radioactive materials for many purposes. Generating electricity in nuclear power stations has been responsible for the leaking of radioactive material into the environment. The levels are usually small, but there have been a number of major incidents around the world, especially at Three Mile Island in the USA in 1979 and at Chernobyl in the Ukraine in 1986. The tsunami and earthquake that caused major damage and loss of life in Japan in 2011 also damaged the Fukoshima nuclear power station resulting in the release of radioactive materials into the air and the ocean as well as making a large area of land around the damaged power station unsafe for humans to live there.

Testing nuclear weapons in the atmosphere has also increased the amounts of radioactive isotopes on the Earth.

Radioactive tracers are used in industry and medicine. Radioactive materials are also used to treat certain forms of cancer. However, the majority of background radiation is natural – the amount produced from medical use in industry is very small indeed.

RADIOACTIVE DECAY

Radioactive decay is a random (unpredictable) process, just like throwing a coin. If we throw a coin we cannot say with certainty whether it will come down heads or tails. If we throw a thousand coins we cannot predict which will land heads and which will land tails. The same is true for radioactive nuclei. It is impossible to tell which nuclei will disintegrate (break down) at any particular time. However, if we threw a thousand coins we would be surprised if the number that landed as heads was not around 500. We know that a normal coin has an equal chance of landing as a head or a tail, so if we got 600 heads we would think it was unusual. If the proportion of heads were much greater than this we would be right to think that the coin was not fair.

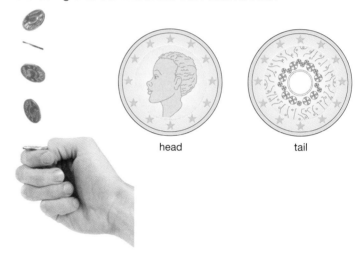

head   tail

▲ Figure 23.4 Throwing a coin

EXPERIMENTAL DEMONSTRATION OF NUCLEAR DECAY

We could, if we had the time, take 1000 coins and throw them. We could then remove all the coins that came down heads, note the number of coins remaining and then repeat the process. If we did this for, say, six trials we would begin to see the trend. A set of typical results is shown in the following table and in Figure 23.5.

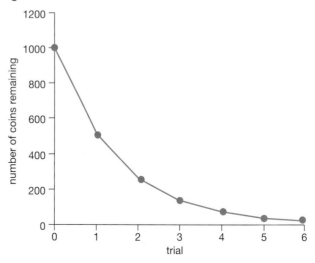

▲ Figure 23.5 Coin-throwing experiment. Each time the coins are thrown about 50% of them land as 'heads' and are removed from the pile. The graph decreases steeply at first but then does so more and more slowly.

Trial	Number of coins remaining
0	1000
1	519
2	264
3	140
4	72
5	33
6	19

EXTENSION WORK

The coin-throwing experiment is a model of radioactive decay. A good model will show the features of the real process. We must remember that models have limitations, however, and do not perfectly represent the actual process. One limitation of the experiment is that of scale. In just 1 g of uranium-235 there are millions of nuclei, and our model uses just 1000 coins. The model would be better if we used, say, 1 000 000 coins, but would take too long to perform. If we use a computer model to throw the coins we could deal with more realistic numbers.

Notice that the graph in Figure 23.5 falls steeply at first and more slowly after each throw. How quickly the graph falls depends on how many heads occur on each throw. But as the number of coins decreases, the number of coins that come up heads also gets smaller. This graph follows a rule: the smaller the quantity, the more slowly the quantity decreases. The quantity here is the number of coins still in the experiment. The name for this kind of decrease proportional to size is called exponential decay.

If we have a sample of a radioactive material, it will contain millions of atoms. The process of decay is random, so we don't know when an atom will decay but there will be a probability that a certain fraction of them will disintegrate in a particular time. This is the same as in the coin toss – there was a 50% probability that the coins would land heads each time we conducted a trial. Once an unstable nucleus has disintegrated, it is out of the game – it won't be around to disintegrate during the next period of time. If we plot a graph of number of disintegrations per second against time for a radioactive isotope we would, therefore, expect the rate of decay to fall as time passes because there are fewer nuclei to decay.

HALF-LIFE

Our coin-tossing model of radioactive decay shows a graph that approaches the horizontal axis more and more slowly as time passes. The model will produce a number of throws after which all the coins have been taken out of the game. The number is likely to vary from trial to trial because the model becomes less and less reliable as the number of coins becomes smaller. With real radioactive decay we use a measure of activity called the half-life. This is defined as follows.

The half-life of a radioactive sample is the average time taken for half the original mass of the sample to decay. If the amount of radioactive matter has halved then the activity of the decay halves – this activity is what is measured in finding the half-life of an isotope. The half-life is different for different radioactive isotopes.

Figure 23.6 shows what this means. After one half-life period, $t_{\frac{1}{2}}$, the amount of the original unstable element has halved. After a second period of time, $t_{\frac{1}{2}}$, the amount has halved again, and so on.

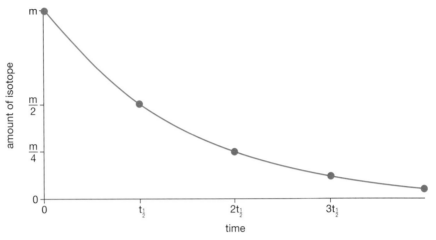

▲ Figure 23.6 Graph showing the half-life period for a radioactive isotope

MEASURING THE HALF-LIFE OF A RADIOACTIVE ISOTOPE

To measure the half-life of a radioactive material (radioisotope) we must measure the activity of the sample at regular times. This is done using a Geiger–Müller (GM) tube linked to a rate meter. Before taking measurements from the sample, we must measure the local background radiation. We must subtract the background radiation measurement from measurements taken from the sample so we know the radiation produced by the sample itself. We then measure the rate of decay of the sample at regular time intervals. The rate of decay is shown by the count rate on the rate meter. The results should be recorded in a table like the one shown below.

Average background radiation measured over 5 min = x Bq		
Time, t/min	Count rate/Bq	Corrected count rate, C/Bq
0	y_0	$y_0 - x$
5	y_5	$y_5 - x$

The rate of decay, C, corrected for background radiation, is proportional to the amount of radioactive isotope present. If we plot a graph of C against time, t, we can measure the half-life from the graph, as shown in Figure 23.7.

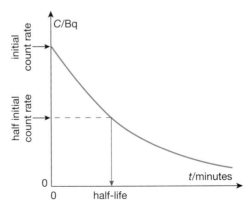

▲ Figure 23.7 You can find the half-life by reading from the graph the time taken for the count rate to halve.

As we have already mentioned, different isotopes can have very different half-lives. Some examples of different half-lives are shown in the table below.

Isotope	Half-life	Decay process
uranium-238	4.5 billion years	α particle emission
radium-226	1590 years	α particle emission,　γ ray emission
radon-222	3.825 days	α particle emission

Isotopes with short half-lives are suited to medical use (see page 241). This is because the activity of a source will rapidly become very small as the isotope decays quickly.

Isotopes used for dating samples of organic material need to have very long half-lives. This is because the activity will become difficult to measure accurately if it drops below a certain level. In Chapter 24, we shall see that there are suitable isotopes for these different applications.

HALF-LIFE CALCULATIONS

Graphs of activity, in becquerels, against time can be used to find the half-life of an isotope, and this half-life information can be used to make predictions of the activity of the radioisotope at a later time.

EXAMPLE 1

The activity of a sample of a certain isotope is found to be 200 Bq.

If the isotope has a half-life of 20 minutes, what will the activity of the sample be after one hour?

After 20 minutes the activity will have halved to 100 Bq.

After 40 minutes (two half-lives) it will have halved again to 50 Bq. After 60 minutes it will have halved again, so the activity will be 25 Bq.

What is the level of activity of this sample after three hours?

Three hours = 9 × 20 minutes – that is, nine half-life periods. This means the activity will have halved nine times. The level of activity (and the amount of the radioisotope remaining) will be:

$$\frac{1}{2} \times \frac{1}{2} \times \frac{1}{2} \times \frac{1}{2} \times \frac{1}{2} \times \frac{1}{2} \times \frac{1}{2} \times \frac{1}{2} \times \frac{1}{2}$$

or $\frac{1}{2^9}$ of the original value, and so $\frac{1}{512}$ of the original activity or amount.

CHAPTER QUESTIONS

SKILLS　CRITICAL THINKING

More questions on radioactive decay can be found at the end of Unit 7 on page 255.

1 a Explain what is meant by background radiation.

　b Explain the difference between natural background radiation and artificial background radiation.

　c Give three different sources of background radiation. Say whether your examples are natural or artificial sources.

2 a Explain simply the principle of the Geiger–Müller tube.

b The Geiger–Müller tube is often connected to a rate meter. Explain what this instrument measures.

c The rate meter is calibrated in kBq. How is this unit defined?

3 a Define what is meant by the half-life of a radioactive material.

b Radioactive decay is a random process. Explain what this means.

4 The activity of a radioactive sample is measured. The activity, corrected for background radiation, is found to be 240 Bq. The activity is measured again after 1 hour 30 minutes and is now 30 Bq. What is the half-life of the sample?

5 In another model of radioactive decay, a student fills a burette with water, as shown in the diagram, and starts a timer at the instant the tap at the bottom is opened. She notes the height of the column of water at regular intervals. It takes 35 seconds to empty from 50 ml to 25 ml. Assuming that the arrangement provides a good model of radioactive decay:

a How long will it take for three quarters of the water in the burette to drain away?

b How much water should remain in the burette after 1¾ minutes?

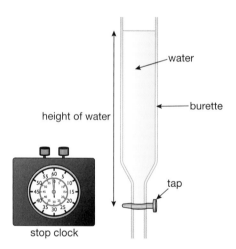

stop clock

6 A student wants to measure the half-life of a radioactive isotope. He is told the isotope has a half-life of between 10 and 20 minutes. Illustrating your answers as appropriate, describe:

a the measurements that he should take

b how he should use the measurements to arrive at an estimate of the half-life for the isotope.

24 APPLICATIONS OF RADIOACTIVITY

Radioactivity has a wide variety of uses, including medicine, industry and power generation. In this chapter, you will read about these uses and also learn about the dangers associated with the use of radioactivity.

LEARNING OBJECTIVES

- Describe uses of radioactivity in industry and medicine
- Describe the difference between contamination and irradiation
- Be able to describe the dangers of ionising radiations including:

- that radiation can cause mutations in living organisms
- that radiation can damage cells and tissue
- the problems arising from the disposal of radioactive waste and how the associated risks can be reduced

THE USE OF RADIOACTIVITY IN MEDICINE

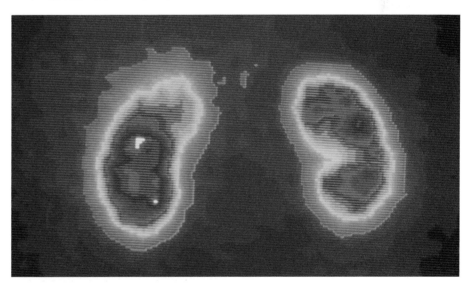

▲ Figure 24.1 This scan shows the kidneys in a patient's body.

USING TRACERS IN DIAGNOSIS

Radioactive isotopes are used as tracers to help doctors identify diseased organs (like the kidneys or the liver). A radioactive tracer is a chemical compound that emits gamma radiation. The tracer is taken orally (swallowed) by the patient or injected. Its journey around the body can then be traced (followed) using a gamma ray camera.

Different compounds are chosen for different diagnostic tasks. For example, the isotope iodine-123 is absorbed by the thyroid gland (a part of the body found in the neck that controls growth) in the same way as the stable form of iodine. The isotope decays and emits gamma radiation. A gamma ray camera can then be used to form a clear image of the thyroid gland.

The half-life of iodine-123 is about 13 hours. A short half-life is important as this means that the activity of the tracer decreases to a very low level in a few days.

Other isotopes are used to image specific parts of the body. Technetium-99 is the most widely used isotope in medical imaging. It is used to help identify medical problems that affect many parts of the body. Figure 24.1 shows a scan of a patient's kidneys. It shows clearly that one of the kidneys is not working properly.

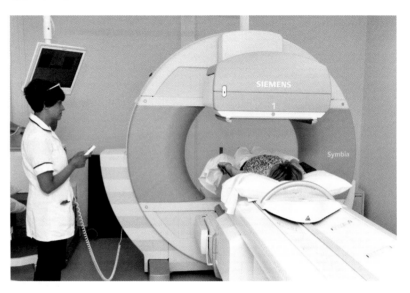

▲ Figure 24.2 Scanner used to provide 3D images of a patient's body

Imaging techniques enable doctors to produce three-dimensional (3D) computer images of parts of a patient's body. These are of great value in diagnosis. Figure 24.2 shows the kind of equipment used for three-dimensional imaging.

TREATMENT

Radiation from isotopes can have various effects on the cells that make up our bodies. Low doses of radiation may have no lasting effect. Higher doses may cause the cells to stop working properly as the radiation damages the DNA in the cells. This can lead to abnormal growth and cancer. Very high doses will kill living cells.

Cancer can be treated by surgery that involves cutting out cancerous cells. Another way of treating cancer is to kill the cancer cells inside the body. This can be done with chemicals containing radioactive isotopes. Unfortunately, the radiation kills healthy cells as well as diseased ones. To reduce the damage to healthy tissue, chemicals are used to target the location of the cancer in the body. They may emit either alpha or beta radiation. Both these types of radiation have a short range in the body, so they will affect only a small volume of tissue close to the target.

The radioisotope iodine-131 is used in the treatment of various diseases of the thyroid gland. It has a half-life of about eight days and decays by beta particle emission.

STERILISATION USING RADIATION

Gamma radiation can kill bacteria and viruses. It is therefore used to kill these microorganisms on surgical instruments and other medical equipment. The technique is called irradiation. The items to be sterilised are placed in secure bags to ensure that they cannot be re-contaminated before use. The gamma radiation will pass through the packaging and destroy bacteria without damaging the item.

Irradiation will not destroy any poisons that bacteria may have already produced in the food before it is treated.

Irradiation does not destroy vitamins in the food like other means of killing bacteria, such as high-temperature treatment.

Some food products are treated in a similar way to make sure that they are free from any bacteria that will cause the food to rot or will cause food poisoning. The irradiation of food is an issue that causes concern amongst the public and is not a widely used procedure at the present time.

Irradiation such as the deliberate exposure of food products and surgical instruments to controlled amounts of radiation should not be confused with radioactive contamination. If radioactive waste is accidentally released either into the air or the sea it could result in fish, animals or agricultural crops being contaminated with radioactive material.

▲ Figure 24.3 Gamma radiation is used to sterilise medical equipment.

THE USE OF RADIOACTIVITY IN INDUSTRY

GAMMA RADIOGRAPHY

▲ Figure 24.4 A gamma camera image used to view inside a valve

A gamma ray camera is like the x-ray cameras used to examine the contents of your luggage at airports. A source of gamma radiation is placed on one side of the object to be scanned and a gamma camera is placed on the other. Gamma rays pass through more objects than x-rays. They can be used to check for faults in casting (making things out of metal) or welding (joining metal objects together). Without the technique of gamma radiography, neither problem could be detected unless the welding or casting were cut through. An additional advantage of gamma radiography over the use of x-rays for this purpose is that gamma sources can be small and do not require a power source or large equipment.

GAUGING

In industrial processes, raw materials and fuel are stored in large tanks called hoppers. Figure 24.5 shows how radioactive isotopes are used to gauge, or measure, how much material there is in a storage container.

The coal absorbs a large amount of the radiation so the reading on the lower detector will be small. As the upper part of the hopper is empty the upper detector will have a high reading.

This method of gauging has several advantages over other methods. There is no contact with the material being gauged. Also, coal dust might cause false readings with an optical gauging system (one using light beams). Coal dust is much less dense than coal so the gamma ray system still works properly.

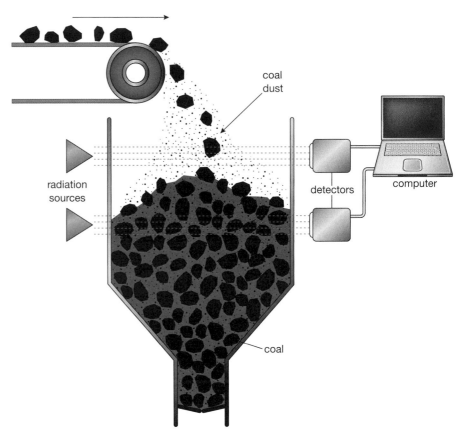

▲ Figure 24.5 The amount of coal in the hopper can be measured using gamma radiation.

Another example of gauging uses a similar process to monitor the thickness of plastic sheeting and film. The thicker the sheet, the greater the amount of radiation it absorbs and the amount passing through gets smaller. By measuring the amount of radiation that passes through the sheeting, its thickness can be closely controlled during manufacture.

TRACING AND MEASURING THE FLOW OF LIQUIDS AND GASES

Radioisotopes are used to check the flow of liquids in industrial processes. Very tiny amounts of radiation can easily be detected. Complex piping systems, like those used in in power stations, can be monitored for leaks. Radioactive tracers are even used to measure the rate of spread of sewage (human waste) (Figure 24.6)!

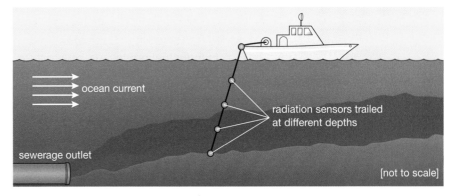

▲ Figure 24.6 Radioactive tracers released with the sewage allow its spread to be monitored to make sure the concentration does not reach harmful levels in any area.

RADIOACTIVE DATING

A variety of different methods involving radioisotopes are used to date minerals and organic matter. The most widely known method is radiocarbon dating. This is used to find the age of organic matter – for example, from trees and animals – that was once living. We shall also look at techniques that are used to find the age of inorganic material like rocks and minerals.

RADIOCARBON DATING

Radiocarbon dating measures the level of an isotope called carbon-14 (C-14). This is made in the atmosphere. Cosmic rays from space are continually raining down upon the Earth. These have a lot of energy. When they hit atoms of gas in the upper layers of the atmosphere, the nuclei of the atoms break apart. The parts fly off at high speed. If they hit other atoms they can cause nuclear transformations (changes) to take place. These transformations turn the elements in the air into different isotopes. One such collision involves a fast-moving neutron striking an atom of nitrogen. (Nitrogen forms nearly 80% of our atmosphere.) The nuclear equation for this process is:

$$^{14}_{7}N + ^{1}_{0}n \rightarrow ^{14}_{6}C + ^{1}_{1}p$$

Notice that, as in the other nuclear equations we have seen, the top numbers – which show the number of nucleons – add up to the same total on each side of the equation. This is because the mass number is conserved. The bottom numbers – which show the amount of charge on the particles – are also conserved.

You have seen the notation for neutrons and protons in Chapter 22. A reminder of neutron notation is shown in Figure 24.7, and of proton notation in Figure 24.8.

The result of the collision of a neutron with a nitrogen atom is a nuclear transformation. The nitrogen atom is transformed into an atom of the radioactive isotope of carbon, carbon-14.

As we have already mentioned, isotopes of an element have the same chemical behaviour. This means that the carbon-14 atoms react with oxygen in our atmosphere to form carbon dioxide, just like the much more common and stable isotope, carbon-12. The carbon dioxide is then absorbed by plants in the process of **photosynthesis**. As a result, a proportion of the carbon that makes up any plant will be the radioactive form, carbon-14. Included in plant material, the radioactive isotope carbon-14 enters the food chain, which means that animals and humans will also have a proportion of carbon-14 in their bodies. These carbon-14 atoms will decay but, in living plants and animals, they are continuously replaced by new ones.

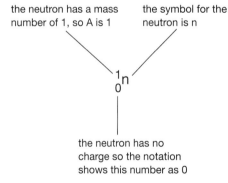

the neutron has a mass number of 1, so A is 1 — the symbol for the neutron is n

$^{1}_{0}n$

the neutron has no charge so the notation shows this number as 0

▲ Figure 24.7 The neutron

When an organism (living thing) dies, the replacement process stops. As time passes, the radioactive carbon decays and the proportion of radioactive carbon in the remains of the plant or animal, compared with the stable carbon isotope, decreases.

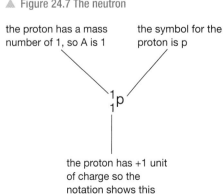

the proton has a mass number of 1, so A is 1 — the symbol for the proton is p

$^{1}_{1}p$

the proton has +1 unit of charge so the notation shows this number as 1

▲ Figure 24.8 The proton

The half-life for the decay of carbon-14 is approximately 5600 years. This means that every 5600 years the proportion of carbon-14 in dead plant and animal material will halve. The amount of carbon-14 present in a sample of dead plant or animal material is found by measuring the activity of the sample. This is compared with the amount of carbon-14 that would have been present when the sample was part of a living organism. From this, it is possible to estimate when the source of the sample died.

EXAMPLE 1

120 g of living wood has a radioactive activity (corrected for background radiation) of 24 Bq. A 120 g sample of wood from an historical site is found to have an activity of 6 Bq. If the half-life of carbon-14 is 5600 years, estimate the age of the wood from the site.

$$6 \text{ Bq} = 24 \text{ Bq} \times \frac{1}{2} \times \frac{1}{2}$$

Since the activity of the sample has halved twice from that expected in living wood, two half-lives must have passed.

The age of the sample is therefore around 2 × 5600 years = 11 200 years.

There are limitations to the method of radiocarbon dating. It assumes the level of cosmic radiation reaching the Earth is constant, which is not necessarily accurate. Fortunately, the technique has been adjusted to take the variations of cosmic ray activity into account. This is done by testing samples of a known age, like material from the mummies (preserved corpses) of Egyptian Pharaohs and from very ancient living trees.

The radiocarbon method is not used to date samples older than 50 000–60 000 years because, after 10 half-lives, the amount of carbon-14 remaining in samples is too small to measure accurately.

EXTENSION WORK

Radiocarbon dating may be used as a context in questions about the half-life of radioactive isotopes. What follows on dating the age of rocks is included for interest only.

DATING THE AGE OF ROCKS

Inorganic, non-living matter does not absorb carbon-14, so different techniques must be used for finding out the age of rocks and minerals.

When a radioactive substance decays it transforms into a different isotope, sometimes of the same element, sometimes of a different element. The original radioisotope is called the parent nuclide (unstable nucleus) and the product is called the daughter nuclide. Many of the products of decay, the daughter isotopes, are also unstable and these too decay in turn. This means that as the parent isotope decays it breeds a whole family of elements in what is called a decay series. The end of the decay series is a stable isotope – one that does not decay further.

The table shows some radioactive parent isotopes with the stable daughters formed at the end of their particular decay series. The half-life quoted is the time for half the original number of parent nuclei to decay to the stable daughter element.

Radioactive parent isotope	Stable daughter element	Half-life/years
potassium-40	argon-40	1.25 billion
thorium-232	lead-208	14 billion
uranium-235	lead-207	704 billion
uranium-238	lead-206	4.47 billion
carbon-14	nitrogen-14	5568

For rocks containing such radioactive isotopes, the proportion of parent to stable daughter nuclide gives a measure of the age of the rock. Notice that the half-lives of most of the radioactive parent isotopes are extremely long, in some cases greater than the lifetime of the Earth.

The decay series of potassium-40 ends with argon gas. As potassium-40 decays in igneous rock, the argon produced remains trapped in the rock. Igneous rocks are formed when molten rock becomes solid. Igneous rocks are non-porous. The proportion of argon to potassium-40 again gives a measure of the age of the rock.

THE DANGERS TO HEALTH OF IONISING RADIATION

Ionising radiation can damage the molecules that make up the cells of living tissue. Cells suffer this kind of damage all the time for many different reasons. Fortunately, cells can repair or replace themselves given time so, usually, no permanent damage results. However, if cells suffer repeated damage because of ionising radiation, the cell may be killed. Alternatively the cell may start to behave in an unexpected way because it has been damaged. We call this effect cell mutation. Some types of cancer happen because damaged cells start to divide uncontrollably and no longer perform their correct function.

Different types of ionising radiation present different risks. Alpha particles have the greatest ionising effect, but they cannot pass through many materials. This means that an alpha source presents little risk, as alpha particles do not penetrate (pass through) the skin. The problem of alpha radiation is much greater if the source of alpha particles is taken into the body. Here the radiation will be very close to many different types of cells and they may be damaged if the exposure is prolonged. Alpha emitters can be breathed in or taken in through eating food. Radon gas is a decay product of radium and is an alpha emitter. It therefore presents a real risk to health. Smokers greatly increase their exposure to this kind of damage as they draw the radiation source right into their lungs (cigarette smoke contains radon).

Beta and gamma radiation do provide a serious health risk when outside the body. Both can penetrate skin and flesh (body) and can cause cell damage by ionisation. Gamma radiation, as we have mentioned earlier, is the most penetrating. The damage caused by gamma rays will depend on how much of their energy is absorbed by ionising atoms along their path. Beta and gamma emitters that are absorbed by the body present less risk than alpha emitters, because of their lower ionising power.

In all cases, the longer the period of exposure to radiation the greater the risk of serious cell damage. Workers in the nuclear industry wear badges to indicate their level of exposure. Some are pieces of photographic film that become increasingly 'foggy' (unclear) as the radiation exposure increases. Another type of badge uses a property called thermoluminescence. Thermoluminescence means that the exposed material will give out light when it is warmed. The radiation releases energy to make heat so the thermoluminescent badges give out more light when exposed to higher levels of radiation. Workers have their badges checked regularly and this gives a measure of their overall exposure to radiation.

SAFE HANDLING OF RADIOACTIVE MATERIALS

Samples of radioactive isotopes used in schools and colleges are very small. This is to limit the risk to users, particularly those who use them regularly – the teachers! Although the risk is small, certain precautions must be followed. The samples are stored in lead-lined containers to block even the most penetrating form of radiation, gamma rays. The containers are clearly labelled as a radiation hazard (danger) and must be stored in a locked metal cabinet. The samples are handled using tongs and are kept as far from the body as possible.

In the nuclear industry and research laboratories, much larger amounts of radioactive material are used. These have to be handled with great care. Very energetic sources will be handled remotely by operators who are protected by lead shields, concrete and thick glass viewing panels. As has been mentioned earlier, neutron radiation is absorbed by lighter elements and waste materials, like spent uranium fuel rods from nuclear reactors, are stored under water until the neutron radiation levels drop to a safe level.

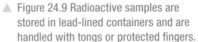

▲ Figure 24.9 Radioactive samples are stored in lead-lined containers and are handled with tongs or protected fingers.

▲ Figure 24.10 Industrial sources of radioactivity must be handed with a lot of care.

The major problem with nuclear materials is long-term storage. Some materials have extremely long half-lives so they remain active for thousands and sometimes tens of thousands of years. Nuclear waste must be stored in sealed containers that must be capable of containing the radioactivity for enormously long periods of time.

CHAPTER QUESTIONS

More questions on applications of radioactivity can be found at the end of Unit 7 on page 255.

SKILLS REASONING

1 The most widely used isotope in medicine is technetium-99m. It has a half-life of six hours and decays by the emission of low-energy gamma rays and beta particles.

SKILLS CREATIVITY, REASONING

a Explain why the characteristics of technetium-99m make it suitable for diagnostic use in medicine.

b Technetium-99m can be chemically attached to a wide variety of pharmaceutical products so that it can be targeted at particular tissues or organs. How can its progress through the body be measured and monitored?

SKILLS REASONING

2 Technetium-99m is produced from molybdenum-99 in a device called a technetium generator. Which decay process – α, β or γ – could cause molybdenum-99 to decay to technetium-99m? Explain your answer.

HINT

Look at Chapter 22.

SKILLS CREATIVITY, REASONING

3 A radioactive isotope of iodine is used in both the diagnosis and treatment of a condition of the thyroid gland. This gland naturally takes up ordinary iodine as part of its function. If a patient has an overactive thyroid it concentrates too much iodine in the gland and this has serious effects on the patient's health.

How might the radioisotope iodine-131 be used to:

a identify an overactive thyroid gland?

b treat the overactive thyroid?

(Iodine-131 has a half-life of eight days and is a high-energy beta emitter.)

SKILLS   CRITICAL THINKING

4 a Explain the difference between radioactive contamination and irradiation.

    b Give an example of a use of irradiation.

SKILLS   CREATIVITY, REASONING

5 Paper is made in a variety of different 'weights', with different thicknesses. How could ionising radiation be used to check the thickness of paper during production? You should consider the following:

■ the type of radiation to be used

■ how it will be used to measure the paper thickness

■ what checks should be made to ensure that the measurements are accurate

■ safety procedures.

SKILLS   CRITICAL THINKING

6 Radiocarbon dating is used to estimate the age of organic (once-living) materials. It uses a radioisotope of carbon, carbon-14 (C-14).

    a How is carbon-14 formed?

    b Why does all living matter contain a proportion of C-14?

    c What happens to the proportion of C-14 in an organism once it has died?

    d What assumptions are made in the process of radiocarbon dating?

    e Why is this method unsuitable for accurately dating material that is more than 50 000 years old?

SKILLS   PROBLEM SOLVING

7 Most radioactive isotopes of elements have half-lives that are extremely short compared to the age of the Earth. The Earth is about 4.5 billion years old. Radium has a half-life of only about 1600 years.

    a How many half-lives of radium have there been since the Earth formed?

SKILLS   ANALYSIS

Student A says all the radium formed when the Earth condensed out of the Sun's atmosphere, should have decayed away to an unmeasurably small amount by now.

Student B says that it depends on how much there was to start with.

Student C says that there is still a significant amount of radium on the Earth.

    b Which student or students are correct? Give reasons for your answer.

SKILLS   REASONING

8 An isotope that decays by alpha emission is relatively safe when outside the body but very dangerous if absorbed by the body, either through breathing or eating.

    a Explain why this is so.

    b Why is radon-220 a particularly dangerous isotope?

25 FISSION AND FUSION

Scientists have speculated about the nature of the atom for thousands of years, but it is only relatively recently that our current ideas were developed. In this chapter you will read about how our ideas about the structure of the atom have developed over the centuries and how we use nuclear energy to produce electricity.

LEARNING OBJECTIVES

- Know that nuclear reactions, including fission, fusion and radioactive decay, can be a source of energy

- Understand how a nucleus of U-235 can be split (the process of fission) by collision with a neutron, and that this process releases energy as kinetic energy of the fission products

- Know that the fission of U-235 produces two radioactive daughter nuclei and a small number of neutrons

- Describe how a chain reaction can be set up if the neutrons produced by one fission strike other U-235 nuclei

- Describe the role played by the control rods and moderator in the fission process

- Understand the role of shielding around a nuclear reactor

- Explain the difference between nuclear fusion and nuclear fission

- Describe nuclear fusion as the creation of larger nuclei resulting in a loss of mass from smaller nuclei, accompanied by a release of energy

- Know that fusion is the energy source for stars

- Explain why nuclear fusion does not happen at low temperatures and pressures, due to electrostatic repulsion of protons

NUCLEAR REACTIONS AS A SOURCE OF ENERGY

Nuclear reactions involve a change in the qualities of atoms. Heavy atoms may split into lighter atoms and other pieces in a process called fission. Lighter atoms may be forced to join together to make heavier atoms in a process called fusion. In either process, the mass of the starting atoms is greater than the mass of the products. This missing mass has been converted into energy.

Within the core of the Earth, radioactive isotopes of elements like uranium, thorium and potassium provide a large proportion of the heat within the Earth itself through radioactive decay.

In the Sun, hydrogen is converted into helium in a **fusion reaction** providing us with a continuous supply of energy in the form of heat and other electromagnetic radiation.

NUCLEAR FISSION

Uranium-235 is used as fuel in a nuclear reactor. It is used because its nuclei can be split by a neutron. The process of splitting an atom is called fission.

Uranium-235 is called a fissile material because it goes through the splitting process easily. The fission process is shown in Figure 25.1.

In the fission reaction, a slow-moving neutron is absorbed by a nucleus of uranium-235.

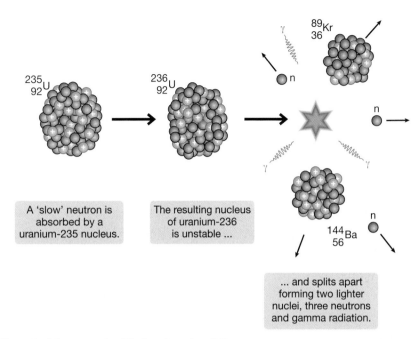

A 'slow' neutron is absorbed by a uranium-235 nucleus.

The resulting nucleus of uranium-236 is unstable ...

... and splits apart forming two lighter nuclei, three neutrons and gamma radiation.

▲ Figure 25.1 One example of fission of uranium-235

The resulting nucleus of uranium-236 is unstable and splits apart. The fragments of this decay are the two daughter nuclei of barium-144 and krypton-89. The decay also produces gamma radiation and three more neutrons. The equation for this decay is:

$$^{236}_{92}\text{U} \rightarrow {}^{144}_{56}\text{Ba} + {}^{89}_{36}\text{Kr} + 3{}^{1}_{0}\text{n} + \gamma \text{ radiation}$$

The fission reaction produces a huge amount of energy. This is because the mass of the products, the barium and krypton nuclei and the three neutrons, is slightly less than that of the original uranium-236 nuclei. This lost mass is converted to energy. Most of the energy is carried away as the kinetic energy of the two lighter nuclei. Some is emitted as gamma radiation. The three neutrons produced by the fission may hit other nuclei of uranium-235, so causing the process to repeat, as shown in Figure 25.2. If one neutron from each fission causes one nearby uranium-235 to split, then the fission reaction will keep going. If more than one neutron from each fission causes fission in surrounding nuclei, then the reaction gets faster and faster – a bit like an avalanche.

EXTENSION WORK

A 'slow' neutron is a low-energy neutron produced by a nuclear decay. Faster moving, more energetic neutrons do not cause fission.

This is called a chain reaction. Each fission results in more nuclei splitting apart. If the amount of uranium-235 is small, many of the neutrons released do not hit other uranium nuclei and the reaction does not get faster and faster. For a chain reaction to happen there must be a minimum amount of the uranium-235. This minimum amount is called the critical mass.

In an atomic bomb two pieces of fissile material (isotopes that can be triggered into splitting apart) that are smaller than the critical mass are forced together under high pressure to form a mass greater than the critical mass. The result is a chain reaction with the rapid and uncontrolled release of huge amounts of energy.

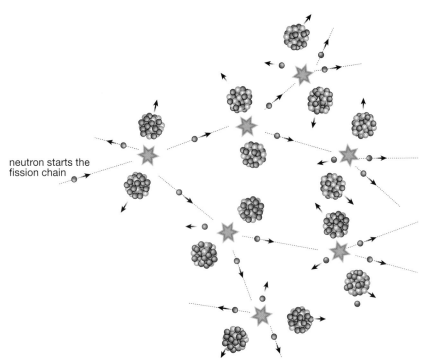

▲ Figure 25.2 A chain reaction in uranium-235

If this is allowed to take place in a nuclear reactor, the reactor core overheats, resulting in a nuclear explosion with the sudden release of enormous amounts of heat energy and radiation. In a nuclear reactor the process is controlled so that the heat energy is released over a longer period of time. The heat produced in the core or heart of the reactor is used to heat water. The steam produced then drives turbines (engines) to turn generators. The basic parts of a nuclear reactor are shown in Figure 25.3.

The reactor core contains fuel rods of enriched uranium. Enriched uranium is uranium-238 with a higher proportion of uranium-235 than is found in natural reserves of uranium. Graphite is used as a **moderator**. The job of

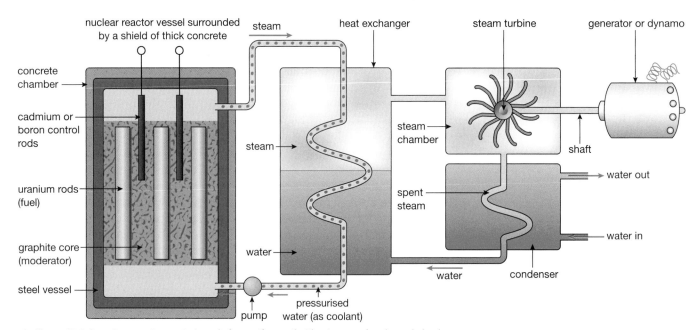

▲ Figure 25.3 A nuclear reactor controls a chain reaction so that heat energy is released slowly.

the moderator is to absorb some of the kinetic energy of the neutrons to slow them down. This is because slow neutrons are more easily absorbed by uranium-235. A neutron slowed in this way can start the fission process.

In the nuclear reactor there are also control rods, made of boron or cadmium. These absorb the neutrons and take them out of the fission process completely. When the control rods are fully inside the core, the chain reaction is almost completely stopped and the rate of production of heat is low. As the control rods are withdrawn, the rate of fission increases producing heat at a greater rate.

The reactor vessel is made of steel and surrounded by a concrete layer about 5 metres in thickness. This prevents any radiation escaping, even neutrons.

The nuclear process in a reactor produces a variety of different types of radioactive material. Some have relatively short half-lives and decay rapidly. These soon become safe to handle and do not present problems of long-term storage. Other materials have extremely long half-lives. These will continue to produce dangerous levels of ionising radiation for thousands of years. These waste products present a serious problem for long-term storage. They are usually sealed (closed) in containers which are then buried deep underground. The sites for underground storage have to be carefully selected. The rock must be water resistant and the geology of the site must be stable – storing waste in earthquake zones or areas of volcanic activity would not be sensible.

Some reactors are designed to produce plutonium. Plutonium is a very radioactive artificial element. Small amounts of plutonium represent a serious danger to health. Plutonium is another fissile material. If a large enough mass of plutonium is brought together a chain reaction will start. Plutonium can be used in the production of nuclear weapons.

Nuclear power stations do not produce carbon dioxide or acidic gases as fossil fuel power stations do. This means that nuclear power does not contribute to global warming or acid rain. Only small amounts of uranium are needed for a chain reaction and the supply of nuclear fuel will last many hundreds of years – unlike some fossil fuels that could run out in the next fifty years.

NUCLEAR FUSION

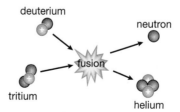

▲ Figure 25.4 Here a nucleus of deuterium collides with a nucleus of tritium. They undergo fusion to form the nucleus of helium, a neutron and a large amount of energy.

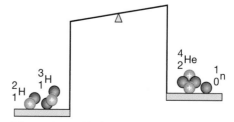

▲ Figure 25.5 The mass of the products of fusion is smaller than the two hydrogen nuclei.

Figure 25.4 shows a fusion process that is being used in projects to develop nuclear fusion reactors. Two isotopes of hydrogen, deuterium (H-2) and tritium (H-3), collide at very high speed. The result is the formation of a helium nucleus, a neutron and a large amount of energy. The fusion process is the energy source of our Sun and all stars.

This process uses materials that are more readily available than the uranium used in a conventional fission reactor and has the advantage of producing no radioactive waste. The problem is creating the very high temperatures needed to make the deuterium and tritium nuclei collide with enough energy to overcome the repulsive (pushing away) force between the positive electric charge in the nuclei of each isotope. The high temperature (100 million °C) liquid must be contained within a very strong magnetic field or 'magnetic bottle'. In order to increase the chance of fusion between the light nuclei in the hot liquid, the pressure within the liquid must be very high.

CHAPTER QUESTIONS

More questions on the structure of the atom can be found at the end of Unit 7 on page 255.

 SKILLS CRITICAL THINKING

1 a Uranium-235 (U-235) is a fissile material. What does this mean?

b If there is a large enough mass of U-235, it may cause a chain reaction. (This is called the critical mass for the isotope.)

i Describe how a chain reaction may take place when a U-235 nucleus splits apart.

ii A chain reaction is unlikely to take place if there is only a small amount of U-235. Explain why this is so.

2 List two advantages and two disadvantages of nuclear fission as a way of producing energy.

 SKILLS REASONING

3 In a nuclear fission reactor, fuel rods made up of a fissile material, like U-235, are lowered into tubes in a large block of graphite. In between the fuel rods are control rods made of boron which may be raised out of the reactor or lowered into it. The reactor container (vessel) is surrounded by steel and a very thick layer of concrete.

a Explain the purpose of the graphite around the fuel rods.

b Describe what effect the action of raising the control rods out of the reactor core has on the behaviour of the reactor.

c Explain how the workers in a nuclear power station are protected from the dangerous types of radiation produced by this nuclear reactor.

 SKILLS CRITICAL THINKING

4 Describe the differences between nuclear fission and nuclear fusion.

5 Explain why creating energy in a nuclear reactor by fusion is so much more difficult than creating power by nuclear fission.

UNIT QUESTIONS

SKILLS   PROBLEM SOLVING   **1**   **a** Which of the following particles has the smallest mass?

    A alpha

    B neutron

    C proton

    D electron

    (1)

SKILLS   CRITICAL THINKING   **b** When a radioactive isotope emits an alpha particle its

    A atomic mass decreases and its atomic number decreases

    B atomic mass increases and its atomic number decreases

    C atomic mass decreases and its atomic number increases

    D atomic mass is unchanged and its atomic number decreases

    (1)

    c The emission of which of the following particles from a radioisotope causes it to change into the next element in the periodic table?

    A positron

    B electron

    C neutron

    D alpha

    (1)

SKILLS   PROBLEM SOLVING   **d** A radioactive isotope has an activity of 1000 Bq; after 5 hours its activity has fallen to 125 Bq. The half-life of this isotope is

    A 1 hour

    B 75 minutes

    C 125 minutes

    D 2.5 hours

    (1)

SKILLS   CRITICAL THINKING   **e** Which of the following particles that may be emitted in radioactive decay is not ionising?

    A alpha

    B beta minus

    C neutron

    D beta plus

    (1)

(Total for Question 1 = 5 marks)

2

In the Sun and other stars the temperature and pressure make hydrogen nuclei fuse together to form heavier nuclei. This process produces a very large amount of energy, various types of radiation and helium. The steps of this fusion process are shown in the following nuclear equations.

Balance these equations by filling in the boxes:

a $^1_1H + {}^1_1H \rightarrow {}^{\square}_1H + {}^0_{\square}e$ **(2)**

b $^2_1H + {}^1_1H \rightarrow {}^{\square}_{\square}He$ **(2)**

c $^{\square}_2He + {}^{\square}_2He \rightarrow {}^4_2He + {}^1_{\square}\square + {}^1_{\square}\square$ **(3)**

(Total for Question 2 = 7 marks)

 3

Here are descriptions of some nuclear particles:

Particle A has 0 mass and a charge +1.

Particle B has 1 mass and a charge 0.

Particle C has 1 mass and a charge +1.

Particle D has 0 mass and a charge −1.

Particle E has 4 mass and a charge +2.

a State which particle is an alpha. **(1)**

b State which particle is a beta minus. **(1)**

c State which particle is a positron. **(1)**

d State which particle is a proton. **(1)**

e State which particle is a neutron. **(1)**

(Total for Question 3 = 5 marks)

4

Two types of nuclear reactor are used to produce energy. Fission reactors are already in use in nuclear power plants around the world. Fusion reactors are still at the experimental stage.

a Explain what is meant by:

  i nuclear fission **(4)**

  ii nuclear fusion **(4)**

b In a fission reactor the process of generating energy is a chain reaction using a suitable fissile fuel like uranium-235.

  i Explain what a chain reaction is, and **(4)**

  ii how it be controlled or shut down completely. **(4)**

(Total for Question 4 = 16 marks)

5

a Explain the conditions that are needed to cause nuclear fusion. **(3)**

b State two reasons why nuclear fusion reactors may be a better alternative to nuclear fission reactors. **(2)**

(Total for Question 5 = 5 marks)

6

a Explain what the letters A and Z tell us about the structure (make up) of the nucleus of any element X.

$$^A_Z X$$

(2)

b Here is a list of the type of radiation that may be emitted (given out) by an unstable nucleus:

A alpha   **B** beta   **C** gamma   **D** neutron

  i   Which type affects A but not Z? **(1)**

  ii   Which type affects Z but not A? **(1)**

  iii   Which type affects both A and Z? **(1)**

  iv   Which type affects neither A nor Z? **(1)**

(Total for Question 6 = 6 marks)

UNIT 8
ASTROPHYSICS

A supernova is an exploding star. The explosion occurs when the gravitational forces within a star are so great that its core collapses releasing huge amounts of energy. It is the largest explosion to take place in space. In this section we will see how important gravitational forces are in our Universe.

26 MOTION IN THE UNIVERSE

We live on a planet called Earth. It is one of many planets that, together with their moons, form our Solar System. Gravitational forces hold our Solar System together. These forces cause the planets, asteroids and comets to orbit the Sun, and moons and artificial satellites to orbit the planets. In this chapter we will look at the key role played by gravitational forces in controlling these movements.

UNITS

In this section you will need to use kilogram (kg) as the unit of mass, metre (m) as the unit of length, metre/second (m/s) as the unit of speed, metre/second² (m/s²) as the unit of acceleration, newton (N) as the unit of force, second (s) as the unit of time and newton/kilogram (N/kg) as the unit of gravitational field strength.

▲ Figure 26.1 The force of gravity keeps these moons in orbit around Jupiter.

LEARNING OBJECTIVES

- Explain that gravitational force causes moons to orbit planets; causes the planets and comets to orbit the Sun; causes artificial satellites to orbit the Earth

- Understand why gravitational field strength, g, varies and know that it is different on other planets and the Moon from that on the Earth

- Describe the differences in the orbits of comets, moons and planets

- Use the relationship between orbital speed, orbital radius and time period:

$$v = 2 \times \pi \times \frac{\text{orbital radius, } r}{\text{time period, } T}$$

THE SOLAR SYSTEM

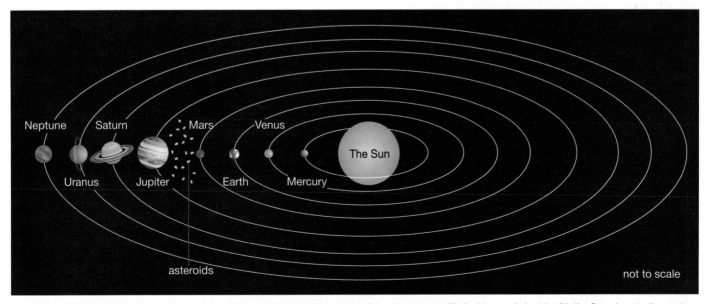

Neptune   Saturn   Mars   Venus   The Sun
Uranus   Jupiter   Earth   Mercury
asteroids   not to scale

▲ Figure 26.2 The Earth is one of eight planets that orbit the Sun. The orbits of the planets are elliptical (an oval shape) with the Sun close to the centre.

WHY DO OBJECTS MOVE IN A CIRCLE?

Figure 26.3 shows a boy swinging a heavy ball around on a wire. To make this ball travel in a circle he needs to spin around and at the same time pull on the wire. Without this continuous pulling force the ball will not travel in a circle.

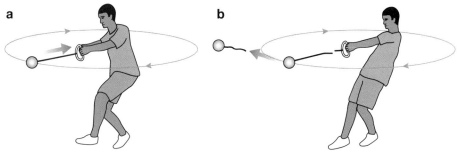

▲ Figure 26.3 **a** A 'pulling' force has to be applied to the ball to make it travel in a circle. **b** If the wire breaks or the boy releases the handle the ball flies away.

Planets and comets travel around the Sun. Moons and satellites travel around the planets. For this to happen there must be forces being applied to them. There is no string or wire to pull on as in the example above, so where do these forces come from? In 1687, Isaac Newton suggested his theory of gravity to explain these movements.

Newton suggested that between any two objects there is always a force of attraction. This attraction is due to the masses of the objects. He called this force gravitational force.

He suggested that the size of this force depends on the:

1 masses of the two objects

2 distance between the masses.

The greater the masses of the two objects the stronger the attractive forces between them.

If the distance between the masses is increased the forces between them decrease.

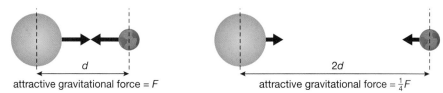

▲ Figure 26.4 Gravitational forces obey an inverse square law – that is, if the distance between the masses is doubled, the forces between them are **quartered**; if the distance between them is **trebled**, the forces become one ninth of what they were.

The gravitational attraction between two objects with small masses is tiny. Only when one or both of the objects has a very large mass – for example, a moon or a planet – is the force of attraction obvious.

Our Sun is massive. It contains over 99% of the mass of the Solar System. It is the gravitational attraction between this mass and each of the planets that holds the Solar System together and causes the planets to follow their curved paths.

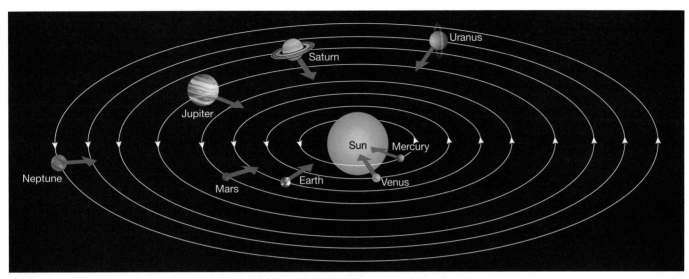

▲ Figure 26.5 Gravitational forces make the planets follow nearly circular paths.

Those planets that are closest to the Sun feel the greatest attraction and so follow the most curved paths. Planets that are the furthest from the Sun feel the weakest pull and follow the least curved path.

EXTENSION WORK

The planet Neptune is a very long way from the Sun so the gravitational force between them is very small. This means that the orbit of Neptune is not very curved and it takes a very long time for it to complete one orbit.

Planet	Average distance from Sun compared with the Earth	Time for one orbit of the Sun in Earth years
Mercury	0.4	0.2
Venus	0.7	0.6
Earth	1.0	1.0
Mars	1.5	1.9
Jupiter	5.0	12
Saturn	9.5	30
Uranus	19	84
Neptune	30	165

▲ Figure 26.6 Some of the many moons of Saturn.

SATELLITES AND MOONS

A satellite is an object that orbits a planet. There are two types of satellite: natural and artificial (human-made).

Natural satellites are called moons. The Earth has just one moon. It is the fifth largest moon in our Solar System, approximately 340 000 km from Earth, and takes just over 27 days to complete one orbit. Although we call our moon 'The Moon' it is not unique. Many planets have moons. Some have more than one, for example, Mars has two moons while Jupiter and Saturn have more than 60 each. All moons have circular orbits because of the gravitational forces between them and their planet.

▲ Figure 26.7 This satellite is held in orbit by the gravitational attraction between it and the Earth.

Since the late 1950s humans have been able to launch, and to put into orbit around the Earth, objects like the one seen in Figure 26.7. These are known as artificial satellites and are extremely useful. Some satellites are put into a very high orbit above the Earth and are used to help us communicate over large distances, for example, for international phone calls or video links, the internet and so on. Some satellites are put into a much lower orbit and are used to monitor in detail the Earth's surface, such as the temperature of the world's oceans or the progress of forest fires.

COMETS

Comets are large rock-like pieces of ice that orbit the Sun. They have very elliptical (elongated) orbits which at times take them very close to the Sun. At other times they travel close to the very edge of our Solar System.

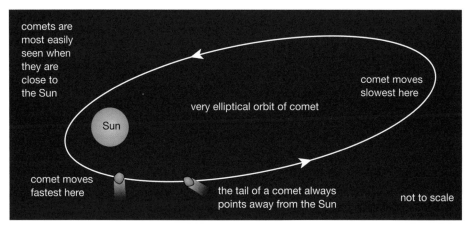

▲ Figure 26.8 The speed of a comet varies enormously. They travel at their fastest when they are very close to the Sun.

▲ Figure 26.9 Halley's comet could be last seen from Earth in 1986.

GRAVITATIONAL FIELD STRENGTH

The strength of gravity on a planet or moon is called its gravitational field strength, and given the symbol g. Different planets have different masses and different radii – both of these will affect their gravitational field strengths.

■ The larger the mass of a planet the greater its gravitational field strength.

■ The larger the radius of a planet the smaller the gravitational field strength at its surface.

The gravitational field strength on the Earth is approximately 10 N/kg whilst on the Moon it is approximately 1.6 N/kg.

Planet	Diameter compared with the Earth	Mass compared with the Earth	Gravitational field strength/N/kg
Mercury	0.4	0.06	4
Venus	0.9	0.38	9
Earth	1.0	1.0	10
Mars	0.5	0.10	4
Jupiter	11	320	23
Saturn	9	95	9
Uranus	4	15	9
Neptune	4	17.0	11

On the Moon, the gravitational field strength is only one sixth that of the Earth's. This means that on the Moon you would weigh six times less. If however you visited Jupiter you would weigh almost 2½ times more!

▲ Figure 26.10 The difference in one person's weight on the Moon and on Jupiter

ORBITAL SPEEDS OF PLANETS AND SATELLITES

The orbital speeds of satellites vary greatly depending on the tasks they are performing. For example, communication satellites are put in high orbits and travel at approximately 3 km/s, while those observing the whole surface of the Earth are put into low polar orbits with speeds of about 8 km/s.

We can calculate the speed of a satellite using the equation:

$$\text{speed, } v = \frac{\text{distance moved, } s}{\text{time taken, } t}$$

The distance moved is the circumference of the circular orbit, r:

distance moved = $2 \times \pi \times$ orbital radius, r

The time period, T, is the time for one complete orbit.

So:

$$\text{orbital speed, } v = \frac{2 \times \pi \times \text{orbital radius}, r}{\text{time period}, T}$$

$$v = \frac{2 \times \pi \times r}{T}$$

The same equation can be used to calculate the speeds of the planets, where r is the average distance from the Sun.

EXAMPLE 1

Calculate the speed of a satellite that is orbiting 200 km above the Earth's surface and completes one orbit in 1 h 24 min. The radius of the Earth is 6400 km.

The radius of the orbit is measured from the centre of the Earth – that is, it is the radius of the Earth plus the height of the orbit above the Earth's surface.

radius of orbit, r = 200 km + 6400 km

$\qquad$ = 6600 km

$\qquad$ = 6 600 000 m

orbital period, T = (60 min + 24 min) × 60 seconds

$\qquad$ = 5040 seconds

orbital speed, $v = \dfrac{2 \times \pi \times r}{T}$

$\qquad = \dfrac{2 \times \pi \times 6\,600\,000 \text{ m}}{5040 \text{ s}}$

$\qquad = \dfrac{41\,500\,000 \text{ m}}{5040 \text{ s}}$

$\qquad$ = 8200 m/s or 8.2 km/s

CHAPTER QUESTIONS

More questions on motion in the Universe can be found at the end of Unit 8 on page 275.

SKILLS CRITICAL THINKING

1 a What is the name of the force that keeps all the planets of the Solar System in orbit around the Sun?

   b What two factors determine the size of this force on the surface of a planet.

   c Describe one difference between the orbits of Mercury and Neptune. Give one reason for this difference.

   d Describe how the speed of a comet changes as it orbits the Sun.

2 Describe the differences between the orbits of a moon, a planet and a comet.

SKILLS PROBLEM SOLVING

3 A satellite is in an orbit 35 786 km above the surface of the Earth. The radius of the Earth is approximately 6400 km. The satellite is moving at 3.07 km/s.

   a How long does the satellite take to orbit the Earth once?

SKILLS REASONING

   b What do you notice about the time period of this satellite?

SKILLS PROBLEM SOLVING

4 If the Earth is approximately 150 million km from the Sun calculate the speed at which the following planets are moving around the Sun. The information you need is given in the table on page 261.

   a Earth $\qquad\qquad$ b Jupiter

27 STELLAR EVOLUTION

The Universe is mainly empty space within which are scattered large numbers of galaxies. Astronomers (scientists who study space and the Universe) believe that there are billions of galaxies in the Universe. The distances between galaxies are millions of times greater than the distances between stars within a galaxy. The distances between the stars in a galaxy are millions of times greater than the distances between planets and the Sun.

▲ Figure 27.1 Photograph of galaxies of the Universe taken by the Hubble space telescope

LEARNING OBJECTIVES

- Know that the Sun is a star at the centre of our Solar System
- Know that the Universe is a large collection of billions of galaxies; galaxies are a large collection of billions of stars; our Solar System is in the Milky Way galaxy
- Understand how stars can be classified according to their colour
- Know that a star's colour is related to its surface temperature
- Describe the evolution of stars of similar mass to the Sun through the following stages:
  - nebula
  - star (main sequence)
  - red giant
  - white dwarf

- Describe the evolution of stars with a mass larger than the Sun

PHYSICS ONLY

- Understand how the brightness of a star at a standard distance can be represented using absolute magnitude
- Draw the main components (parts) of the Hertzsprung–Russell diagram (HR diagram)

THE MILKY WAY

Our nearest star is the Sun. It is approximately 150 million kilometres from the Earth. Its surface temperature is approximately 6000 °C and temperatures within its core are about 15 000 000 °C. Attractive gravitational forces between stars cause them to group together in enormous groups called galaxies. Galaxies consist of billions of stars. Our galaxy is a **spiral galaxy** called the Milky Way. We are approximately two-thirds of the way out from the centre of our galaxy along one of the arms of the spiral.

▲ Figure 27.2 Our galaxy, the Milky Way

▲ Figure 27.3 Our galaxy takes the shape of a spiral, like the one shown here.

▲ Figure 27.4 The colour of the light emitted by this piece of iron tells us how hot it is.

CLASSIFYING STARS

Even within our galaxy we can see a wide variety of stars (Figure 27.2).

Looking up into a clear night sky, especially if we use a telescope, we can see that stars are not all identical. They have different colours, different levels of brightness and appear to be different sizes. Scientists who study the stars in detail have created classes or star groups based upon these similarities and differences.

Why are some stars white, some red and others yellow? Surprisingly we can find the answer to this question in a traditional blacksmiths. To shape a piece of iron without it breaking the iron must be very hot. A skilled blacksmith will know when the temperature of the iron is high enough simply by taking it out of the furnace (very hot oven) and looking at its colour. If it is glowing white or bright orange it is very hot. If it is a dull red it is much cooler and not ready to be shaped.

In a similar way the colours of stars tell us about their temperatures. A very hot star emits more blue in its spectrum and therefore looks blue, a medium star like our Sun looks yellow and cooler stars appear red.

Using the colours of stars and their surface temperatures, scientists have created seven different groups of stars. These groups are called O, B, A, F, G, K, M. O and B are the hottest stars and K and M the coolest.

Star classification	Surface temperature/K	Colour
O	more than 33 000	blue
B	33 000–10 000	blue-white
A	10 000–7500	white
F	7500–6000	yellow-white
G	6000–5200	yellow
K	5200–3700	orange
M	3700–2000	red

PHYSICS ONLY

The brightness of a star

At first there would seem to be little doubt which of these vehicle headlights is the brightest. But is it really that simple? The cars in the distance might have brighter headlights but they appear to be dimmer because they are further away. Also the brightness of headlights in different makes of cars and lorries could be different!

When we look at the stars in the night sky we have similar problems deciding which are the brightest.

The brightness of a star depends on **a** the distance the star is from the Earth and **b** what the star is made from and the kinds of nuclear reactions that are taking place.

As a result there are three different ways in which astronomers describe the brightness of a star:

1 The apparent brightness or magnitude of a star. This is the easiest method and is simply a measure of how bright a star is as seen from the Earth.

2 The absolute brightness or magnitude. This is a measure of how bright stars would appear if they were all placed the same distance away from the Earth. (This distance that astronomers agreed on, called the standard distance, is equal to 10 parsecs or 32.6 light years). This is often a better measure of brightness because it allows scientists to make comparisons between stars.

3 The luminosity of a star. This measures how much energy in the form of light is emitted from a star's surface every second.

END OF PHYSICS ONLY

▲ Figure 27.5 It is impossible to decide which of these vehicles has the brightest headlights just by looking at the photograph.

THE BIRTH OF A STAR

Dust and gas particles are drawn together by gravitational forces.

Compression (pulling together) of particles causes increases in temperature and nuclear reactions begin.

Unused material around the young star begins to group together.

This material may form the planets, moons and comets that orbit the star.

▲ Figure 27.6 Birth of a solar system

Stars like our Sun are formed from large clouds of dust and gas particles we call nebulae or stellar nebulae. These particles are drawn together over a very long period of time by gravitational forces. The particles are pulled together so tightly that there is a very large increase in temperature and pressure. As a result of this nuclear fusion reactions begin. Hydrogen nuclei join together to make larger nuclei and huge amounts of energy in the form of heat and light are released. This incredibly hot ball of gas is a very young star.

THE LIFE OF A STAR

The appearance of a star changes gradually with time. These changes follow a pattern.

When a star first forms, gravitational forces are pulling particles together. Then when nuclear reactions begin, the high temperatures create forces that try to push the particles apart – that is, make the gases expand. When these two forces are balanced, the star is said to be in its main stable period. A star in this main stable period is referred to as a main sequence star (you will learn why later). This period can last for many millions of years. At the moment our Sun is in this stable period.

Towards the end of this stable period, there are less hydrogen nuclei and eventually the hydrogen fusion reactions stop. Gravitational forces are now the largest forces and compress the star. As the star shrinks in size there is a large increase in temperature. So high that fusion reactions between helium nuclei begin. The energy released by these reactions causes the star to expand to many times its original size. As it expands it becomes a little cooler and more of its light energy is emitted in the red part of the spectrum. The star is changing into a red giant.

Sometime later when most of the helium nuclei have fused (joined) together, new nuclear reactions begin, but now the compressive or squashing forces are larger and the star begins to get smaller or contract. This contraction causes an increase in temperature so the star again changes colour. It now emits more blue and white light. It has changed into a white dwarf star. The matter from which white dwarf stars are made is millions of times more dense than any matter found on the Earth.

Finally as a white dwarf star cools it changes into a cold black dwarf star.

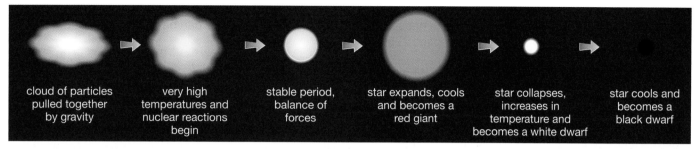

▲ Figure 27.7 The life-cycle of stars with a mass similar to that of our Sun

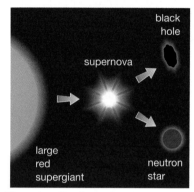

▲ Figure 27.8 The final stages for stars with masses much greater than our Sun

A star that is much larger than our Sun may follow a slightly different path at the end of its life. After the stable period it will expand into a large red supergiant. At the end of this period, as it contracts it becomes unstable. It explodes throwing dust and gas into space to form a new stellar nebula. This exploding star is called a supernova. Any matter remaining will form a very dense neutron star. If the neutron star has a mass that is approximately five times greater than that of our Sun or more, it collapses further to become a black hole.

EXTENSION WORK

The gravitational field of a black hole is so strong that even light is unable to escape from it. This is why they are called 'black holes'.

All the elements in the periodic table up to iron are formed in stars but elements heavier than these are only formed in supernova explosions.

PHYSICS ONLY

The Hertzsprung–Russell diagram (HR diagram)

Two scientists, Ejnar Hertzsprung and Henry Norris Russell, separately discovered the relationship between the brightness of a star and its surface temperature. They created a diagram that shows the relationship between the brightness, temperature and classification of a star.

The importance of this diagram is that by finding a star's position in the diagram we then know its internal structure and at what stage it is at in its life.

For example, Sirius lies in the main sequence belt so it is a star similar to our Sun and fusion reactions between hydrogen nuclei are taking place in its core. In the next stage of its life it will become a red giant.

Rigel is in the supergiants region of the diagram and so has already passed through the main sequence part of its life. The majority of fusion reactions taking place in its core are between helium nuclei. It will eventually become a neutron star or a black hole.

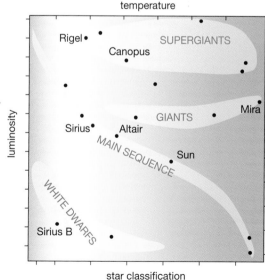

▲ Figure 27.9 The Hertzsprung–Russell diagram is important because it can be used to describe the life story of a star and to predict the final stages of its life.

END OF PHYSICS ONLY

CHAPTER QUESTIONS

More questions on stars and their life cycles can be found at the end of Unit 8 on page 275.

SKILLS CRITICAL THINKING

1 a What do we call a large group of stars?

  b Why do these groups form?

  c What is the name of the group where we live?

  d How many other groups of stars do we think exist in the Universe?

SKILLS ANALYSIS

2 Using the table on page 267 answer the following questions.

  a Describe two properties of stars that belong in classes B and K.

  b To which group does a star belong if its surface temperature is 2500 K?

  c To which two groups might a star belong if it is emitting a lot of yellow light?

  d What range of surface temperatures might this star have?

SKILLS CRITICAL THINKING

6

3 Explain the difference between the apparent and absolute brightness of a star.

4

4 Describe the different stages of a life of a star that is much more massive than our Sun.

5 a What is a supernova?

  b Under what conditions does a supernova occur?

SKILLS ANALYSIS

5

6 In Figure 27.9 you can see two stars named Canopus and Sirius B.

  a In which parts of the HR diagram are these stars?

SKILLS REASONING

6

  b Describe the reactions that are taking place in the core of Canopus.

  c Describe the next stage in the life of Sirius B.

28 COSMOLOGY

How did the Universe begin? Did it have a beginning or did it always exist? Will it have an end or will it go on for ever? As scientists discover more and more about the Universe, they believe they are getting closer and closer to discovering the answers to these questions. The first clues to solving these problems came from observing explosions like the one shown in Figure 28.1.

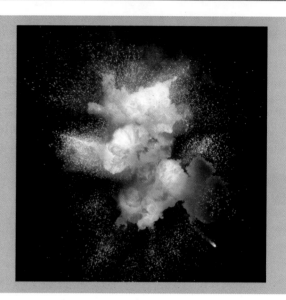

▶ Figure 28.1 This photograph was taken a fraction of a second after an explosion. The movement of the particles and the energy released have some similarities to events we can see in space. Could this be a clue that suggests that the Universe started with a 'Big Bang'?

LEARNING OBJECTIVES

- Describe the past evolution of the Universe and the main arguments in favour of the Big Bang

- Describe evidence that supports the Big Bang theory (red-shift and cosmic microwave background (CMB) radiation)

- Describe that if a wave source is moving relative to an observer there will be a change in the observed frequency and wavelength

- Use the equation relating change in wavelength, wavelength, velocity of a galaxy and the speed of light:

$$\frac{\text{change in wavelength}}{\text{reference wavelength}} = \frac{\text{velocity of a galaxy}}{\text{speed of light}}$$

$$\frac{\lambda - \lambda_0}{\lambda_0} = \frac{\Delta \lambda}{\lambda_0} = \frac{v}{c}$$

- Describe the red-shift in light received from galaxies at different distances from the Earth

- Explain why the red-shift of galaxies provides evidence for the expansion of the Universe

▲ Figure 28.2 A telescope to look at stars

THE DOPPLER EFFECT

If we look at the stars and galaxies in our Universe using a simple optical telescope it is impossible to tell if they are moving, and if they are moving then how fast or in which direction. If however we look at the spectrum of light emitted by a star or galaxy we can discover all of the above. We can do this by using an effect called the Doppler effect (see page 104), first described by an Austrian scientist named Christian Doppler in 1842.

Before we look at the Doppler effect in space, let us look at some more familiar situations where we can observe it.

If we close our eyes and listen to the engine of a fast moving car it is usually very simple to decide if it is approaching us or moving away. The loudness of the motor will of course change but so too will the frequency of the sound we hear. As the car approaches the frequency of the sound produced by its

approaching and departing

As the car approaches, the pitch is higher.

As the car leaves, the pitch gets lower.

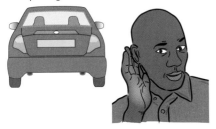

▲ Figure 28.3 An everyday example of the Doppler effect

motor appears to be high but when it moves away it appears to have a lower frequency. This apparent change in frequency is called the Doppler effect. The faster the car is moving the greater this change in frequency.

The Doppler effect is a property of all waves. It happens not only with sound waves but also with light waves. When astronomers look at light spectra from distant stars and galaxies they can see the Doppler effect.

Figure 28.4 shows the spectra from four different objects. The first is the spectrum from our Sun. The dark lines we can see are called **absorption** lines. They are the frequencies of light that have been absorbed by hydrogen. The second spectrum is from a nearby galaxy. Here we can clearly see that the positions of the dark lines have moved towards the red part of the spectrum. The third is the spectrum from a very distant galaxy – the lines have moved even further towards the red. The last spectrum shows the greatest shift for the furthest galaxy. We call this displacement '**red-shift**'. Like the changes in the frequency of the car engine it indicates a relative motion between the galaxies and the observer.

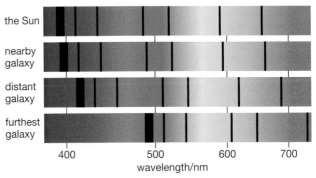

▲ Figure 28.4 Light spectra for four different objects

Red-shift indicates that the source of the light waves is moving away from the observer. Blue-shift would indicate that the source of light is moving towards the observer. We can see why this is the case in Figure 28.5.

a light from a stationary (not moving) star, for example, our Sun

b light from a similar star moving away from the Earth is stretched and so has a longer wavelength – that is, we see red-shift

c light from a similar star moving towards the Earth is compressed and so has a shorter wavelength – that is, we see blue-shift

▲ Figure 28.5 Demonstrations of light from a star moving away from Earth and moving towards it

When we compare the light emitted from all the different galaxies a clear pattern emerges. Almost all the galaxies emit light with red-shift. The further away a galaxy is the greater the red-shift and therefore the faster it is moving away from us. We see exactly this pattern in situations that involve explosions.

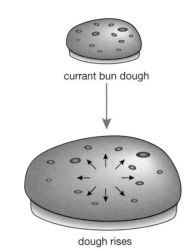

currant bun dough

dough rises

▲ Figure 28.6 As the bun expands the currants get further apart. Similarly as the Universe expands the galaxies get further apart.

So scientists now believe that the Universe is expanding and that at some time in the past, all the matter in the Universe was in one place just before an explosion. This theory is called the Big Bang theory.

Another way to imagine the expansion of the Universe is to observe what happens to the dough and currants in a bun when they are placed in an oven to cook. Over a period of time the dough expands and the currants all move away from each other. In this model the dough represents the Universe, increasing in size as time passes. The currants represent galaxies that are seen to be moving away from each other as the Universe expands.

THE DOPPLER EQUATION

The equation below shows us how to calculate the speed at which a star or galaxy is moving relative to us.

$$\frac{\text{change in wavelength, } \Delta\lambda}{\text{reference wavelength, } \lambda_0} = \frac{\text{velocity of a galaxy, } v}{\text{speed of light, } c}$$

$$\frac{\lambda - \lambda_0}{\lambda_0} = \frac{\Delta\lambda}{\lambda_0} = \frac{v}{c}$$

EXAMPLE 1

Light emitted by the Sun has a wavelength of 434 nm. A distant galaxy emits the same light but because of the Doppler effect it has a wavelength of 482 nm. Calculate the speed at which the galaxy is moving relative to the Earth. (speed of light = 3×10^8 m/s)

$$\frac{\Delta\lambda}{\lambda_0} = \frac{v}{c}$$

$$v = \frac{\Delta\lambda}{\lambda_0} \times c$$

$$= \frac{(482 - 434) \times 10^{-9} \text{ m}}{434 \times 10^{-9} \text{ m}} \times 3 \times 10^8 \text{ m/s}$$

$$= \left(\frac{48}{434}\right) \times 3 \times 10^8 \text{ m/s}$$

$$= 0.332 \times 10^8 \text{ m/s}$$

$$= 33\,200 \text{ km/s}$$

Because the apparent wavelength is longer, this is an example of red-shift. The galaxy is therefore moving away from the Earth.

COSMIC MICROWAVE BACKGROUND (CMB) RADIATION

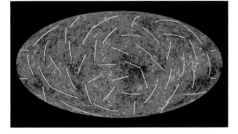

▲ Figure 28.7 A map of the cosmic background radiation, the afterglow of the Big Bang

To reinforce the idea that the Universe began with a Big Bang, scientists searched for something else that we might connect with an explosion – that is, the release of energy. This energy would be in the form of waves. But as the Universe expands these waves would become stretched and therefore they would have much longer wavelengths than before. In fact these wavelengths were predicted to be so long that they would be in the microwave part of the electromagnetic spectrum. In the 1960s scientists detected this afterglow of energy. As was predicted, they are microwaves and can be detected in all directions in the Universe. They became known as cosmic microwave background (CMB) radiation.

LOOKING AHEAD – DARK ENERGY

Having found evidence that the Universe began with a Big Bang and is expanding, scientists are now asking the question, 'Will the Universe continue to expand for ever or will gravity gradually slow it down and reverse the process, pulling all the matter back to a single point (the Big Crunch)?' The rate at which the expansion of the Universe slows depends upon how much mass there is in the Universe. The greater the mass the more the expansion of the Universe will be slowed.

However, in 1988, observations of a very distant galaxy made by the Hubble space telescope showed that the expansion of the Universe had previously been slower. This suggested that the expansion of the Universe is not slowing but accelerating. Scientists are still not able to explain this. They have a theory that there is energy in the Universe that is causing this acceleration. They cannot see it or detect it so it is called 'dark energy'. If you continue to study physics after your International GCSE you will learn more about this.

CHAPTER QUESTIONS

More questions on cosmology can be found at the end of Unit 8 on page 275.

SKILLS REASONING

1 A racing car travelling at high speed passes an observer. Describe the frequency of the sound produced by the car's engine heard by the observer

   a as the car approaches, and

   b as the car moves away.

SKILLS CRITICAL THINKING

2 The compression of light waves emitted by a galaxy that is approaching an observer is called blue-shift. What is the stretching of light waves by a galaxy moving away from an observer called?

SKILLS REASONING

3 How will the light spectra of a distant galaxy differ from that of a very distant galaxy? Explain what this difference indicates.

4 What two observed features of our Universe suggest that it began with a Big Bang?

SKILLS PROBLEM SOLVING

5 The spectrum of a distant galaxy shows a change in position of absorption lines of 300 nm compared with the Sun's spectrum. If these lines in the Sun's spectrum correspond to a light with wavelength of 600 nm calculate the speed of this galaxy relative to the speed of the Earth.
(speed of light = 3×10^8 m/s)

END OF PHYSICS ONLY

UNIT QUESTIONS

 1

a Which of the following statements is incorrect?

  A Gravitational forces keep our Moon in orbit around the Earth.

  B Comets and moons orbit planets.

  C Planets orbit the Sun.

  D Artificial satellites orbit a planet. **(1)**

b Which of the following sequences might be followed by a star that has a mass much larger than our Sun?

  A main sequence, stellar nebula, red giant, white dwarf

  B stellar nebula, main sequence, red supergiant, black hole

  C stellar nebula, main sequence, white dwarf, black hole

  D main sequence, red giant, neutron star, white dwarf **(1)**

c Which are the most common reactions that take place in the core of a star that is in the main sequence?

  A nuclear fission reactions between helium nuclei

  B nuclear fission reactions between hydrogen nuclei

  C nuclear fusion reactions between helium nuclei

  D nuclear fusion reactions between hydrogen nuclei **(1)**

d Which of the following observations does not support the Big Bang theory?

  A cosmic microwave background radiation

  B blue-shift in the spectra of galaxies

  C red-shift in the spectra of galaxies

  D accelerating galaxies **(1)**

(Total for Question 1 = 4 marks)

2 Copy and complete the following passage about gravitational forces, filling in the spaces.

Gravitational forces cause planets and _____ to _____ the Sun. The planets _____ to the Sun, for example, _____ and _____, experience the _____ forces and so have the most _____ orbits. The planets _____ from the Sun, for example, _____ and _____, experience the _____ forces and so have the least _____ orbits. The shape of a planet's orbit is not quite _____. It is _____.

All objects that orbit a planet are called _____ but natural ones are called_____.

(Total for Question 2 = 16 marks)

 3

A man has a mass of 80kg. On the Earth he weighs 800N. Using the table on page 263 calculate the weight of this man on the surface of:

a Venus **(3)**

b Mars **(3)**

c Neptune. **(3)**

(Total for Question 3 = 9 marks)

SKILLS PROBLEM SOLVING **4** Calculate the speed of a satellite that is orbiting 250 km above the Earth's surface and completes one orbit in 2 hr.

(Total for Question 4 = 5 marks)

SKILLS ANALYSIS **5** The table below contains information about some of the planets in our Solar System.

Planet	Surface gravity compared with the Earth	Distance from Sun compared with the Earth	Period/Earth years
Mercury	0.4	0.4	0.2
Venus	0.9	0.7	0.6
Mars	0.4	1.5	1.9
Jupiter	2.6	5.0	12
Saturn	1.1	9.5	30

a Name three planets that have a weaker gravitational pull on their surface than there is on Earth. **(1)**

HINT

circumference of a circle = $2\pi r$

b How long is a year on Saturn? **(1)**

SKILLS PROBLEM SOLVING c If the distance from the Earth to the Sun is 150 million kilometers, calculate the distance of Saturn from the Sun. **(2)**

d Assuming that the orbital path for Saturn is circular, calculate its orbital speed in km/s. **(3)**

(Total for Question 5 = 7 marks)

SKILLS INTERPRETATION **6** a Draw a diagram to show the shape of the orbit of a comet around the Sun. **(2)**

SKILLS REASONING b Describe and explain how the speed of a comet changes as it travels around its orbit. **(4)**

(Total for Question 6 = 6 marks)

SKILLS INTERPRETATION **7** Draw the Hertzsprung–Russell diagram. Label the main features and explain why the diagram is useful for astronomers.

((Total for Question 7 = 12 marks)

8 What is cosmic microwave background radiation?

(Total for Question 8 = 3 marks)

SKILLS PROBLEM SOLVING **9** The spectra of two galaxies, A and B, each show a change in the position of absorption lines of 200 nm and 450 nm compared with the Sun's spectrum. If these lines correspond to light with a wavelength of 500 nm in the Sun's spectrum calculate the speed of these galaxies relative to the speed of the Earth. (speed of light = 3×10^8 m/s)

(Total for Question 9 = 10 marks)

APPENDICES

APPENDIX A: PERIODIC TABLE

Period	Group 1	2														3	4	5	6	7	0
1							1 H Hydrogen 1														4 He Helium 2
2	7 Li Lithium 3	9 Be Beryllium 4														11 B Boron 5	12 C Carbon 6	14 N Nitrogen 7	16 O Oxygen 8	19 F Fluorine 9	20 Ne Neon 10
3	23 Na Sodium 11	24 Mg Magnesium 12														27 Al Aluminium 13	28 Si Silicon 14	31 P Phosphorus 15	32 S Sulphur 16	35.5 Cl Chlorine 17	40 Ar Argon 18
4	39 K Potassium 19	40 Ca Calcium 20	45 Sc Scandium 21	48 Ti Titanium 22	51 V Vanadium 23	52 Cr Chromium 24	55 Mn Manganese 25	56 Fe Iron 26	59 Co Cobalt 27	59 Ni Nickel 28	63.5 Cu Copper 29	65 Zn Zinc 30				70 Ga Gallium 31	73 Ge Germanium 32	75 As Arsenic 33	79 Se Selenium 34	80 Br Bromine 35	84 Kr Krypton 36
5	85 Rb Rubidium 37	88 Sr Strontium 38	89 Y Yttrium 39	91 Zr Zirconium 40	93 Nb Niobium 41	96 Mo Molybdenum 42	(99) Tc Technetium 43	101 Ru Ruthenium 44	103 Rh Rhodium 45	106 Pd Palladium 46	108 Ag Silver 47	112 Cd Cadmium 48				115 In Indium 49	119 Sn Tin 50	122 Sb Antimony 51	128 Te Tellurium 52	127 I Iodine 53	131 Xe Xenon 54
6	133 Cs Caesium 55	137 Ba Barium 56	139 La Lanthanum 57	178 Hf Hafnium 72	181 Ta Tantalum 73	184 W Tungsten 74	186 Re Rhenium 75	190 Os Osmium 76	192 Ir Iridium 77	195 Pt Platinum 78	197 Au Gold 79	201 Hg Mercury 80				204 Tl Thallium 81	207 Pb Lead 82	209 Bi Bismuth 83	(210) Po Polonium 84	(210) At Astatine 85	(222) Rn Radon 86
7	(223) Fr Francium 87	(226) Ra Radium 88	(227) Ac Actinium 89																		

140 Ce Cerium 58	141 Pr Praseodymium 59	144 Nd Neodymium 60	(147) Pm Promethium 61	150 Sm Samarium 62	152 Eu Europium 63	157 Gd Gadolinium 64	159 Tb Terbium 65	163 Dy Dysprosium 66	165 Ho Holmium 67	167 Er Erbium 68	169 Tm Thulium 69	173 Yb Ytterbium 70	175 Lu Lutetium 71
232 Th Thorium 90	(231) Pa Protoactinium 91	238 U Uranium 92	(237) Np Neptunium 93	(242) Pu Plutonium 94	(243) Am Americium 95	(247) Cm Curium 96	(247) Bk Berkelium 97	(251) Cf Californium 98	(254) Es Einsteinium 99	(253) Fm Fermium 100	(256) Md Mendelevium 101	(254) No Nobelium 102	(257) Lr Lawrencium 103

a = relative atomic mass
X = atomic symbol
b = atomic number

a
X
Name
b

(Masses in brackets are the mass numbers of the most stable isotope)

APPENDIX B: FORMULAE AND RELATIONSHIPS

1 the relationship between average speed, distance moved and time taken:

$$\text{average speed} = \frac{\text{distance moved}}{\text{time taken}}$$

2 the relationship between force, mass and acceleration:

force = mass × acceleration

3 the relationship between acceleration, change in velocity and time taken:

$$\text{acceleration} = \frac{\text{change in velocity}}{\text{time taken}}$$

4 the relationship between momentum, mass and velocity:

momentum = mass × velocity

momentum = $m \times v$

5 the relationship between density, mass and volume:

$$\text{density} = \frac{\text{mass}}{\text{volume}}$$

6 the relationship between work done, force and distance moved:

work done = force × distance moved

7 the energy relationships:

energy transferred = work done

kinetic energy = $\frac{1}{2}$ × mass × speed²

gravitational potential energy = mass × g × height

8 the relationship between mass, weight and gravitational field strength:

weight = mass × gravitational field strength

9 the relationship between an applied force, the area over which it acts and the resulting pressure:

$$\text{pressure} = \frac{\text{force}}{\text{area}}$$

10 The relationship between the moment of a force and its perpendicular distance from the pivot:

moment = force × perpendicular distance from the pivot

11 the relationship between charge, current, voltage, resistance and electrical power:

charge = current × time

voltage = current × resistance

electrical power = voltage × current

12 the relationship between speed, frequency and wavelength of wave:

wave speed = frequency × wavelength

13 the relationship between turns and voltage for a transformer:

$$\frac{\text{input (primary) voltage}}{\text{output (secondary) voltage}} = \frac{\text{primary turns}}{\text{secondary turns}}$$

14 the relationship between refractive index, angle of incidence and angle of refraction:

$$n = \frac{\sin i}{\sin r}$$

15 the relationship between refractive index and critical angle:

$$\sin c = \frac{1}{n}$$

16 the relationship for efficiency:

$$\text{efficiency} = \frac{\text{useful energy output}}{\text{total energy input}} \times 100\%$$

17 the relationship for pressure difference:

pressure difference = height × density × gravitational field strength

$$p = h \times p \times g$$

18 input power = output power

$$V_p I_p = V_s I_s$$

for 100% efficiency

Reproduced from Pearson Edexcel International GCSE in Physics – Specification – Issue 1 – June 2016 © Pearson Education Limited 2016

APPENDIX C: PHYSICAL QUANTITIES AND UNITS

Fundamental physical quantities

Physical quantity	Unit(s)*
length	**metre (m)** kilometre (km) centimetre (cm) millimetre (m)
mass	**kilogram (kg)** gram (g) milligram (mg)
time	**second (s)** millisecond (ms)
temperature	**Kelvin (K)** degrees Celsius/°C
current	**ampere** or **amp (A)** milliampere or milliamp (mA)

*SI base units are shown in bold.

Derived quantities and units

Physical quantity	Unit(s)
acceleration	m/s^2
area	cm^2 m^2
density	kg/m^3 g/cm^3
electrical charge	coulomb (C)
electrical current	ampere (A)
energy (work)	joule (J) kilojoule (kJ) megajoule (MJ)
frequency	hertz (Hz) kilohertz (kHz)
force	newton (N)
gravitational field strength	N/kg
moment of a force	Nm Ncm
momentum	kg m/s
potential difference (voltage)	volt (V) millivolt (mV)
power	watt (W) kilowatt (kW) megawatt (MW)
pressure	pascal (Pa or N/m^2) N/cm^2
radioactivity	becquerel (Bq)
resistance	ohm (Ω)
speed/velocity	m/s km/h
volume	cm^3 m^3 litre (l) millilitre (ml)

All the units used in science are defined by the Système International d'Unités: SI units.

You will, of course, find measurements of units made in subdivisions and multiples of the base unit. For example, the subdivisions of the kilogram are grams (g) and milligrams (mg). Metres are subdivided into centimetres (cm) and multiplied into kilometres (km). You will also be familiar with other units for time: minutes, hours, days and years, and so on. You will need to take care to convert units in calculations to the base units of kg, m and s when you meet these subdivisions and multiples.

EXAMPLE

Car speedometers (an instrument in a vehicle that shows how fast it is going) read in km/h. Calculate the equivalent speed in m/s of a car travelling at 40 km/h.

There are 1000 metres in a kilometre so
40 km = 40 000 m.

An hour is 3600 seconds, so the car travels 40 000 m in 3600 s.

In one second the car will travel $\dfrac{40\,000\,m}{3600\,s}$

so 40 km/h ≡ 11.1 m/s

Usain Bolt, current world record holder for the 100 m sprint, averaged 10.43 m/s over his record-breaking 100 m.

APPENDIX D: EXPERIMENTAL AND INVESTIGATIVE SKLLS

Your investigative skills will be tested by some of the questions in your examinations. This section summarises some of the skills you may be asked about.

The examination papers will test your ability to do some of these things:

- describe the method for an investigation
- take measurements
- record results in tables
- plot graphs to show results
- analyse results
- evaluate data.

You will learn these skills during practical work throughout your course. This Appendix provides some reminders of the kinds of skills you need. The main chapters in this book contain descriptions of various experiments and it might be helpful to look back at these as well.

METHOD

The method for a practical investigation is a list of instructions for carrying out the investigation. In an exam you may be given a list of apparatus and asked to describe how to use it to investigate a particular question. When you are writing a method, it is important to describe all the steps in the correct order. The person marking the exam paper cannot watch you doing the experiment – if you do not describe a step they will assume you have forgotten about it.

RELIABLE EVIDENCE

In many of the investigations you have carried out, you will have been asked to repeat readings and use a mean value when plotting a graph or writing your conclusion. By repeating results you can find out how reliable your evidence is (see more on this on page 283), and spot any mistakes. Your method should explain that you will repeat each measurement, and that you will discard any anomalous results (ones that do not fit the pattern) before finding a mean value. You should also explain why you would do this.

For some experiments you do not need to take repeat measurements of the same thing. For example, you might be finding out which type of insulation works best by measuring the temperature of hot water in insulated beakers. In this case, you can measure the temperature of the water every minute for 10 or 20 minutes, and plot a line graph of temperature against time. You will be able to spot any anomalous results because they will not lie on the line formed by most of your readings. You do not need to repeat the whole experiment several times.

However you do need to explain this in your method – the person marking your paper will not just assume you understand why drawing a graph like this is useful.

SAFETY PRECAUTIONS

You may also be asked about safety precautions you would take when carrying out an investigation. The safety precautions depend on the type of investigation, but these are some of the things you should think about:

- if you are using weights, make sure they cannot fall and land on your feet
- make sure any clamp stands are fastened to the bench so they cannot overbalance
- if you are stretching wires or other objects that might break and spring back, wear eye protection
- if you are using electricity, make sure you are using a low-voltage source
- if you are using water, make sure that any spills are mopped up so you do not slip
- if you are projecting (firing) an object such as a ball bearing or an arrow make sure that no one is in the area and that an adequate absorbent target is in place
- if you are describing the use of radioactive materials make sure you know how they should be handled and stored safely
- if you are timing the descent of an object such as a model parachute make sure that any chairs or other furniture are out of the way.

Remember that most investigations and practicals are quite safe, and so safety precautions can also refer to care of the apparatus. For example:

- not stretching a spring too far so that it goes beyond its elastic limit
- if you are using electricity to heat something such as fuse wire use a heatproof mat to protect the bench and do not touch hot wires.

MEASUREMENTS

UNITS

Measurements are a key part of investigations, but they do not mean anything unless you also give the units. A length of '15' could mean 15 mm, 15 cm, 15 m, or even 15 km!

RECORDING RESULTS

Results should always be recorded in a table. The factor that you change in an experiment usually goes in the first column. This table shows the results for an experiment to investigate how the current changes when you change the length of a piece of wire in a circuit.

This is what you are going to be testing, so it goes in the first column.	Length of wire/cm	Current/A	This is what you will be measuring.
	10		Always put in the units.
	20		
These are the lengths of wire you will be testing.	30		
	40		Use a ruler when you draw straight lines.
	50		

Sometimes you need to record measurements in more than one situation. If you wanted to record current in a hot wire and a cold one, your results table might look like this:

Length of wire/cm	Current/A		The heading applies to both of the columns underneath it.
	Hot wire	Cold wire	
10			
20			
30			
40			
50			

DEPENDENT, INDEPENDENT AND CONTROL VARIABLES

In an investigation we often want to find out if one physical quantity is dependent on another physical quantity and if we are able to describe the relationship either on a graph, in words or mathematically.

An example would be to investigate whether the electrical resistance of a piece of wire is dependent on its length. Here the dependent variable would be the resistance of the length of the wire and the independent variable, the one we would change for each measurement of the resistance, would be the length of the wire. We would need to be aware that other physical quantities might affect the resistance of the length of wire, for example, the temperature of the wire. This is an example of a control variable; we would check the temperature of the wire and its surroundings throughout the investigation to make sure it did not change. If the temperature of the wire changed it would make our conclusion about how the resistance of a wire depends on its length invalid.

Can you think of any other variables that might affect the resistance of a length of wire that would need to be controlled (kept the same) during an investigation of how the resistance of a wire depends on its length?

A student wants to find out how the voltage output from a PV (photovoltaic) cell depends on the angle it makes to the direction of light shining on it.

a State which quantity is the dependent variable and which is the independent variable.

b i What other quantities would need to be kept the same during the investigation?
ii What do we call these variable quantities that must be kept constant during this investigation?

GRAPHS

Graphs are a very useful way of presenting results. A graph allows you to find patterns in results and also helps you to spot any results that may not fit the pattern. The graph below shows the results of an experiment to find the length of a spring with different weights on it.

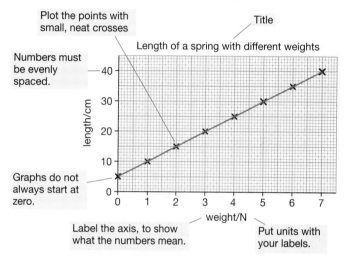

You should normally draw a straight line on a graph that goes through as many points as possible. This is called a line of best fit.

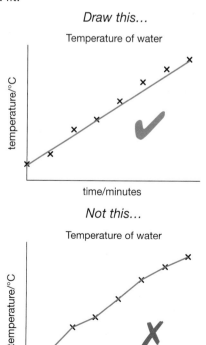

However, a straight line is not always the best way of joining the points. You do need to think about what the results are showing you.

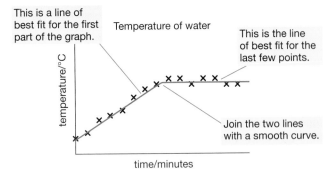

This is a line of best fit for the first part of the graph.

This is the line of best fit for the last few points.

Join the two lines with a smooth curve.

Temperature of water

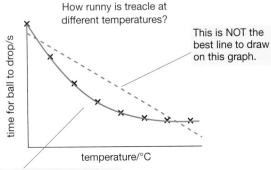

How runny is treacle at different temperatures?

This is NOT the best line to draw on this graph.

Draw a smooth curve through these points. This is a curve of best fit

You can use a graph to spot mistakes – if all of your results except one lie close to either a straight line or a smooth curve, that result is likely to be wrong. You may have made a mistake when taking the measurement, or you may have written it down incorrectly or incorrectly calculated a derived quantity such as acceleration. You should try to work out why that result is wrong, so that you can suggest improvements to your method to improve the accuracy of your results.

SKETCH GRAPHS

You may sometimes be asked to sketch a graph to show the results. A sketch graph shows the relationship between two variables without showing numbers. For example, you might be asked to sketch a graph to show what happens to the current in a wire when you increase the voltage. Your graph might look like the one below.

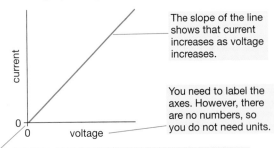

The slope of the line shows that current increases as voltage increases.

You need to label the axes. However, there are no numbers, so you do not need units.

Think about whether the line should go through the origin. In this case, there is no voltage, so the line *should* go through the origin.

You may sometimes need to include units on a sketch graph, if more than one possible scale could be used. For example, if you were asked to sketch a graph to show how the pressure of a fixed volume of gas varies with temperature, either of these graphs would be correct.

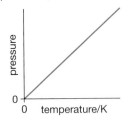
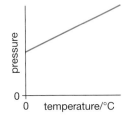

ANALYSING RESULTS

In some investigations you may have to carry out calculations on your results. For example, if you were investigating how the resistance of a wire depended on its length you would need to calculate the resistance from current and voltage measurements. When you show the results of your calculations, you should always give your result to an appropriate number of significant figures. The following examples should make this clear.

The current and voltage in the first table are both measured to two significant figures (2 sf). In this case, it means that the voltage is accurate to the nearest 0.1 V and the current is accurate to the nearest 0.01 A. When you divide the voltage by the current to get the resistance, you get numbers with lots of decimal places. If you just write down the whole answer as displayed on your calculator, this implies that the answer is more accurate than it is. For example, the last column in the table implies that the resistance value is accurate to the nearest 0.000 000 01 Ω and this is obviously not the case!

Voltage/V	Current/A	Resistance/Ω
2.0	0.24	8.33333333
4.0	0.51	7.84313725
6.0	0.85	7.05882353

The results of calculations should be given to the same number of significant figures as in the measurements used to calculate them. If your answer has more significant figures, then you need to round your answer. The table below is a better way of presenting these results.

Voltage/V	Current/A	Resistance/Ω
2.0	0.24	8.3
4.0	0.51	7.8
6.0	0.85	7.1

WRITING A CONCLUSION

You may be asked to write a conclusion for a set of results, or to comment on a conclusion that is given.

For example, this graph shows the results of an experiment to test different insulating materials. The student covered two boiling tubes of water in different materials and measured the temperature every minute.

This is the conclusion the student wrote for the insulating materials investigation:

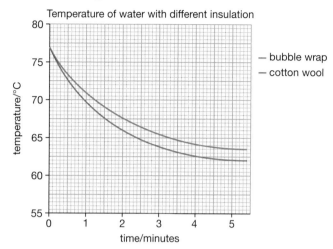

Temperature of water with different insulation

— bubble wrap
— cotton wool

The water in the tube covered in cotton wool cooled from 77 °C to 62 °C, which is 15 °C. The water in the tube covered in bubble wrap cooled from 77 °C to 63.5 °C, which is 13.5 °C. The bubble wrap kept the water warmer, so it is the best insulator.

This conclusion is correct, but it is not the best conclusion that can be drawn from these results. The point about drawing a line of best fit through the points on a graph is that it evens out any errors in measurements. A better way to consider these results is to look at the lines on the graph – the line for cotton wool is steeper, showing that the water in this tube is cooling down faster than the water in the tube covered in bubble wrap.

You may also be asked to compare a conclusion with a prediction made before the investigation. Remember to read what the prediction said, and to clearly say whether or not you think the prediction has been shown to be correct. You should also explain why you think this.

EVALUATION

Evaluation means deciding how accurate and reliable your results are. In the examination, you may be given a set of results and asked to comment on them.

You may be asked to identify any results that do not fit the overall pattern.

You may be asked to comment on how precise or reliable the evidence is, or how to improve the accuracy or reliability of the results.

PRECISION

'Precise' means that things have been measured as accurately as possible. Particularly where the quantities

being measured are small, the measurements should be taken with the most accurate equipment possible. For example, if the experiment described involved measuring very small currents, you might suggest that the results could have been more precise if a milliammeter had been used instead of an ammeter. Just saying something like 'Take more care with the measurements' will not gain any marks!

The accuracy of measurements can be quite poor if human reaction times are involved. For example, if someone is timing how long it takes for a pendulum to complete one swing, they need to start and stop the stop clock as they see the pendulum passing a particular point on its swing. You could suggest two different ways to improve the accuracy of this measurement:

■ Use a light gate and data logger to measure the time – this would remove any errors due to human reaction time.

■ Measure the time it takes the pendulum to complete 10 swings, and then divide the total time by 10 to find the time for one swing. There may still be an error due to human reaction time here, but the error will be spread out over many swings, so the error on the time for one swing will be much smaller.

RELIABILITY

'Reliable' results mean that if the measurements are repeated, the same result should be obtained. You can see how reliable a set of results are by comparing three or more measurements. If the results are close to each other, then the evidence is probably reliable.

The graph of the results of the insulation experiment shows a different way of checking reliability. In this investigation only one set of results was taken for the each type of insulation. However, because the results can be plotted on a line graph, you can use the closeness of the points to the line of best fit on the graph to judge the reliability. In this case, most of the results look reliable, because they are close to the lines of best fit.

MAXIMISING YOUR SUCCESS

The very best way of revising for the exam at the end of your course is to work through past papers set by your examiners, looking especially at the most recent ones. You need to check your answers against the published mark schemes and against the comments in the Examiners' Reports. The Examiners' Reports are particularly useful for these practically-based questions, as what they are looking for isn't always obvious.

This book was written for the new Edexcel International GCSE Physics specification (syllabus) to be examined for the first time in 2019, and so in the early stages there will be a shortage of past papers tied directly to the new specification. In that case, you will have to look instead at the questions set in similar exams in the past.

APPENDIX E: COMMAND WORDS

Command word	Definition
Add/Label	Requires the addition or labelling of a stimulus material given in the question, for example labelling a diagram or adding units to a table.
Calculate	Obtain a numerical answer, showing relevant working.
Comment on	Requires the synthesis of a number of variables from data/information to form a judgement.
Complete	Requires the completion of a table/diagram.
Deduce	Draw/reach conclusion(s) from the information provided.
Describe	To give an account of something. Statements in the response need to be developed, as they are often linked but do not need to include a justification or reason.
Determine	The answer must have an element that is quantitative from the stimulus provided, or must show how the answer can be reached quantitatively. To gain maximum marks, there must be a quantitative element to the answer.
Design	Plan or invent a procedure from existing principles/ideas.
Discuss	Identify the issue/situation/problem/argument that is being assessed within the question. Explore all aspects of an issue/situation/problem/ argument. Investigate the issue/situation etc. by reasoning or argument.
Draw	Produce a diagram either using a ruler or freehand.
Estimate	Find an approximate value, number or quantity from a diagram/given data or through a calculation.
Evaluate	Review information (for example, data, methods) then bring it together to form a conclusion, drawing on evidence including strengths, weaknesses, alternative actions, relevant data or information. Come to a supported judgement of a subject's quality and relate it to its context.

Command word	Definition
Explain	An explanation requires a justification/exemplification of a point. The answer must contain some element of reasoning/justification – this can include mathematical explanations.
Give/State/Name	All of these command words are really synonyms. They generally all require recall of one or more pieces of information.
Give a reason/reasons	When a statement has been made and the requirement is only to give the reason(s) why.
Identify	Usually requires some key information to be selected from a given stimulus/resource.
Justify	Give evidence to support (either the statement given in the question or an earlier answer).
Plot	Produce a graph by marking points accurately on a grid from data that is provided and then draw a line of best fit through these points. A suitable scale and appropriately labelled axes must be included if these are not provided in the question.
Predict	Give an expected result.
Show that	Verify the statement given in the question.
Sketch	Produce a freehand drawing. For a graph, this would need a line and labelled axes with important features indicated. The axes are not scaled.
State what is meant by	When the meaning of a term is expected but there are different ways for how these can be described.
Suggest	Use your knowledge to propose a solution to a problem in a novel context.
Verb proceeding a command word	
Analyse the data/graph to explain	Examine the data/graph in detail to provide an explanation.
Multiple-choice questions	
What, Why	Direct command words used for multiple-choice questions.

GLOSSARY

absorption absorption is the opposite of reflection: soft foam surfaces absorb sound, matt black surfaces absorb light and heat

accelerating getting faster, increasing velocity

acceleration the rate of increasing velocity

alternating voltage a voltage that is continuously changing in value and direction

amplitude the maximum distance moved by a vibrating object from its equilibrium position

angle of incidence angle between the incoming ray and the normal

angle of reflection angle between the reflected ray and the normal

angle of refraction angle between the refracted ray and the normal

apparatus equipment used in investigations and experiments

applications uses

asteroids small rocky objects orbiting the Sun, mostly found between the orbits of Mars and Jupiter.

atoms small particles from which everything is made

attract pull together

bacteria single-celled organisms, some types of which cause illness

calories unit of energy no longer used except in measuring the amount of energy in food

cells sources of electrical energy

circuits in electricity and electronics, complete conducting paths for electricity; circuit is sometimes used as a term for electronic apparatus

collision two or more moving objects hitting each other

comet object often made of ice and rocks that orbits the Sun in an elongated orbit

compressed squeezed into a very small space

compression squashing together

conduction thermal conduction: the movement of heat through a solid; electrical conduction: the movement of electric charge through matter

conductor a material that allows electricity to flow through it easily

control rods rods used in a nuclear reactor to slow down or stop a nuclear chain reaction

convection the movement of heat in a fluid (that is, a gas or liquid) as the fluid expands and rises when warmed

core the centre of something as in, for example, the Earth's core

deceleration getting slower, decreasing velocity

deflected made to change direction

diameter the width of a circle, cylinder or sphere

dimmer switch a device used to alter the brightness of a bulb or light

displacement the distance a particle has moved from its equilibrium (undisturbed) position

dissolves broken down into tiny particles or molecules by the action of a liquid, for example, sugar dissolving in water

Doppler effect the change in frequency (and wavelength) caused by the relative movement of the source of the waves or the observer

drag coefficient a measure of how easily an object can move through a fluid (liquid or gas)

electrical insulators materials that do not allow electric current to pass through them

electrode a metal plate or rod by which electricity can enter or leave an electrical device

electromagnetic spectrum family of waves including radio waves, microwaves, infrared waves, visible light, ultraviolet and x-rays

elements in chemistry: pure substances made up of only one type of atom; in general use: part of something

emission something emitted from a system

emitted given out

evaporate change from a liquid into a gas (usually at a temperature lower than the boiling point of the liquid)

exerted acted on

filament coil of wire in a bulb that glows when electricity passes through it

fission the breaking up of an atom into smaller parts

flex wire

force a push or a pull

fossil fuels substances used to provide energy when burned that have formed over millions of years from dead animal or vegetable matter

fraction a part of; sometimes represented mathematically as a ratio of a smaller number over a larger number, for example, ½ for a half

frequency the number of waves or vibrations made each second

friction a force between two solid surfaces trying to move across each other that tries to stop movement happening

fusion reaction when the centres of atoms (nuclei) join together

generator a device that transfers mechanical energy to electrical energy – that is, a source of electrical energy

glows emits light

gradient the slope of a line or surface

gravitational field strength how great the effect of gravity acting on an object is, usually given as force per unit mass

gravity the force that objects with mass exert on each other

half-life the length of time it takes for the activity of an amount of a radioactive substance to halve

halogen light bulb a light bulb that contains a small amount of a halogen such as iodine or bromine

heating elements coils of wires used to transfer electrical energy to heat energy

hemisphere half a sphere, as in the northern hemisphere referring to the part of the Earth above the equator

hydroelectric power electrical energy produced from the energy in water stored high above ground level in mountain lakes and reservoirs

induced created, caused, produced

inkjet printers printers that create an image by directing droplets of ink onto paper

insulator a material through which it is very difficult or impossible for electricity to flow

inversely proportional something varying such that it decreases by equal amounts as some other factor increases by equal amounts

kinetic energy the type of energy that a moving object stores

light emitting diodes (LEDs) a material that gives off light when current passes through it

light gates electronic systems that are used to switch something on or off, like a digital clock when a light beam between a light source and a detector is broken

longitudinal wave the vibrations of these waves are along the direction in which the wave travels

magnetic field a place where we can detect magnetism

magnification how many times bigger, for example, a magnification of two indicates it is twice as big

mass the amount of matter in an object; that property of an object that determines how easy it is to speed up or slow down

metric related to a system of units of measurement based on kilograms, metres and second, for example, SI units are metric

moderator a material used in nuclear reactors to make neutrons move more slowly

molecules groups of atoms joined together

negatively charged has more negative charges than positive charges

non-porous a material that does not allow liquids to pass through it

non-renewable an energy source that will not last for ever, for example, a fossil fuel like coal will run out eventually

normal (the) a line at 90 degrees to the surface

nuclear to do with the nucleus of an atom, for example, the forces that hold neutrons and protons tightly together in the nucleus is a nuclear force

optical telescope an instrument that we use to see distant object using visible light

orbit the path of a planet around a star or a satellite around a planet

orbital speeds the speeds of objects as they circle other objects, for example, the Moon going around the Earth

ore many chemical elements are bound up with other elements in rocks; these rocks are called ores

oscilloscope a device used to observe waves and vibrations

ozone layer a layer high above the Earth's surface that contains lots of ozone (O_3)

parabola a particular shape of curved surface used to focus light, heat, radio waves, and so on, to a particular point

parallax error not reading a scale on a ruler or other measuring instrument by looking at the scale straight on (at right angles to the scale)

penetrating power the measure of the ability of a wave to travel through objects

perpendicular upright or at right-angles (90 degrees to something)

phenomena events, occurrences

photosynthesis the process in green plants that uses light from the Sun to produce energy for the plant to grow

photovoltaic cells devices that transfer energy from light (typically the Sun) into electrical energy; used in renewable energy projects; not to be confused with solar heat exchangers that transfer heat from the Sun to water, allowing energy to be stored in tanks of heated water

pitch (of a sound) the interpretation of the frequency of a sound by a person's brain/ear

pivot the point around which something can turn or rotate

plugs and sockets the most common way to obtain electrical energy from the mains supply

polarity term used to describe which parts of a magnet are north or south or which parts of an electrical circuit are positive and negative

prisms transparent, triangular shaped pieces of glass or plastic

proportional something varying such that it increases by equal amounts as some other factor increases by equal amounts

quartered decreased by a factor of 4, that is, ¼ of the original

radioactive having an unstable nucleus that will emit particles and waves to achieve a more stable nucleus

radioisotope isotope of an element that has the same number of protons in the nucleus but different numbers of neutrons; a radioisotope is unstable, radioactive; also called radioactive isotope

ratio a way of comparing the quantities of things, for example, a ratio of 3:1 means there is three times the amount of one thing compared to the other

reaction the response to an action; in physics, reaction is the equal force that acts as a result of an action force, equal in size but opposite in direction

rectifier circuits circuits used to change alternating currents into direct currents

red-shift emitted waves have their wavelength increased, that is, moved to the red end of the spectrum

refractive index a measure of the change in speed a wave experiences when it travels across the boundary between two media, e.g. air and glass; the index also describes how much the direction of the wave changes; $n = \sin i/\sin r$

relative charge the charge compared with the charge on an electron

renewable used to describe energy supplies that will not be completely used up in time like the energy from the Sun, wave energy, wind energy, and so on.

repel push away

repulsive forces forces that push away from each other

resistance the difficulty current experiences in a circuit

resultant force the total effect of two or more forces acting on an object; sometimes referred to as an unbalanced force

rotate to turn or spin about an axis (central point)

rotating turning

scanned looked at over an area, as in scanning pages from a book in a photocopier

solenoid a long coil

sonar method of using sound waves to detect objects and measure their distance

spiral galaxy a very large group of stars that forms a spring-like winding shape

squared in maths, something multiplied by itself

sterilised made free of dirt and bacteria

sterilise remove all bacteria and germs

stroboscope apparatus that produces short bright flashes of light at regular intervals (that is at a known frequency, flashes per second); the apparatus allows the frequency to be varied; used to allow a series of images of a moving object to be captured photographically or to determine the speed of rotating objects

subtract take away from

sum the result of adding things together

terminal velocity the maximum speed that a moving object is able to reach

thermal relating to heat; also used to describe upward air currents called convection

thermistor a conductor whose resistance changes a lot when its temperature changes

thrust a type of pushing force, like the thrust of rocket motors

time period time needed to make one complete wave or vibration

tracers the use of substances that can be detected to show the movement of liquids through a person's body or pipes in machinery

transformer device used to change alternating voltages

transverse wave the vibrations of this wave are at right angles to the direction in which the wave is moving

trebled increased by a factor of 3, that is, 3 × the original

tuning fork a steel instrument that when it vibrates produces sound waves of a constant frequency and wavelength

unbalanced force non-zero resultant force

unstable likely to change; referring to an object it means to tip over; in the case of a radioactive isotope it means to emit particles that change the make-up of a nucleus

upthrust an upward force acting on something; an object in a liquid or gas will have an upward force acting on it called the buoyant (to do with floating) upthrust

vacuum containing no matter at all; the space between stars is a nearly perfect vacuum; a vacuum flask has a gap around the inner container that is completely emptied of any matter

vertical upwards, at right angles to the ground

vibrate move continuously back and forth

visible light waves that can be detected by the eye

voltage amount of energy transferred to each coulomb of charge that passes

wavelength the distance from one point on a wave to the same point on the next wave

waves vibrations which carry or move energy

weight the force of gravity acting on a body

weightless the condition in which a body does not experience the force we call weight; if you stand on weighing scale in a lift that is accelerating toward the Earth at 10 m/s^2 the scales will show that you weigh nothing – this is just one way in which weightlessness can happen

INDEX